The Silviculture of Trees used in British Forestry

THE SILVICULTURE OF TREES USED IN BRITISH FORESTRY

by

Peter S. Savill
Lecturer in Silviculture,
University of Oxford

CAB INTERNATIONAL

CAB INTERNATIONAL
Wallingford
Oxon OX10 8DE
UK

Tel: +44 (0)1491 832111
Fax: +44 (0)1491 833508
E-mail: cabi@cabi.org
Telex: 847964 (COMAGG G)

A catalogue record for this book is available from the British Library, London, UK.

ISBN 0 85198 739 7

First printed 1991
Reprinted 1992, 1996, 1998

Printed and bound in the UK by Information Press, Eynsham

CONTENTS

ACKNOWLEDGEMENTS ix

INTRODUCTION 1

ABIES
- ALBA 4
- GRANDIS 5
- PROCERA 7

ACER
- CAMPESTRE 10
- PLATANOIDES 10
- PSEUDOPLATANUS 12

ALNUS
- CORDATA 16
- GLUTINOSA 17
- INCANA 17
- RUBRA 21

BETULA
- PENDULA 22
- PUBESCENS 22

CARPINUS
- BETULUS 25

CASTANEA
- SATIVA 27

CHAMAECYPARIS
- LAWSONIANA 29

CORYLUS
- AVELLANA 31

CUPRESSOCYPARIS
- LEYLANDII 33

EUCALYPTUS 34

FAGUS
SYLVATICA 36

FRAXINUS
EXCELSIOR 41

ILEX
AQUIFOLIUM 45

JUGLANS
REGIA 46

LARIX
DECIDUA 50
KAEMPFERI 52
EUROLEPIS 52

MALUS
SYLVESTRIS 55

NOTHOFAGUS
OBLIQUA 56
PROCERA 56

PICEA
ABIES 59
SITCHENSIS 60

PINUS
CONTORTA 65
MURICATA 67
NIGRA 68
NIGRA var. MARITIMA 71
NIGRA var. NIGRA 73
PEUCE 73
RADIATA 75
STROBUS 76
SYLVESTRIS 76

POPULUS 78

PRUNUS
AVIUM 82

PSEUDOTSUGA
MENZIESII 86

QUERCUS
CERRIS 89
PETRAEA 90
ROBUR 90
RUBRA 97

ROBINIA
PSEUDOACACIA 99

SALIX 101

SEQUOIA
SEMPERVIRENS 103

SEQUOIADENDRON
GIGANTEUM 104

SORBUS
ARIA 106
AUCUPARIA 106
TORMINALIS 106

TAXUS
BACCATA 109

THUJA
PLICATA 111

TILIA
CORDATA 114
PLATYPHYLLOS 114

TSUGA
HETEROPHYLLA 117

ULMUS 119

REFERENCES 121

APPENDIX 130
Table 1. Regressions of crown diameters on stem diameters 131
Table 2. Life expectancy of trees 132
Table 3. Ages of maximum mean volume increment 134
Table 4. Maximum dimensions of trees 136
4a. Important native timber producing species 136
4b. Minor native timber producing species 136
4c. Other native trees 137
4d. Exotic conifers 138
4e. Exotic broadleaves 139

INDEX 140

ACKNOWLEDGEMENTS

My thanks are expressed to many colleagues, and past and present students who have contributed, in various ways, to the preparation of this book by referring me to relevant papers in journals and elsewhere. I am especially grateful to Rosemary Wise who prepared all the drawings.

The figures in Tables 1 and 2, and in Appendix Table 3 are Crown copyright and taken from Forestry Commission publications, and are reproduced with their permission. Information on the life expectancy of trees in Appendix Table 2 is reproduced with the permission of the Arboricultural Association, and that on the maximum dimensions of trees, with permission of William Collins and Sons Ltd.

INTRODUCTION

Many books have been written about trees which grow in Britain. They range from guides for identifying trees. One of the best known is Alan Mitchell's (1978) *A field guide to the trees of Britain and northern Europe*, and his more detailed publications dealing with a much smaller range of common conifers and broadleaves (Mitchell, 1985a and b). Numerous publications by the Forestry Commission (e.g. Hibberd, 1986) also provide some basic information which can be useful when selecting species for particular sites.

Few, however, cater particularly well for the person who requires detailed information about the silvicultural requirements of individual species. Among the more recent authors to do this were M.L. Anderson in his *Selection of tree species*, which was first published in 1950, and Macdonald *et al.* (1957) in their *Exotic forest trees in Great Britain*. The information they provide about species, though still valuable, has inevitably become dated as thinking has changed, and techniques and knowledge improved.

This publication aims to provide an up-to-date guide which can be used when selecting species. The requirements of individual trees are described, for areas in which they are likely to do well. No particular consideration has been given to the relative economics of the different species: the main emphasis is upon the biological suitability of species to sites. It is assumed that the reader is reasonably well informed about the principles of forestry practice. The general considerations which must be taken into account when selecting a tree for planting are described in many silviculture textbooks including *Plantation forestry in temperate regions* by Savill and Evans (1986). The amount of information available about the climatic and site requirements of species, and other features of their silviculture varies considerably but is closely related to how common they are. For some, especially the major conifers, oak and beech, knowledge is quite comprehensive, while for others like lime and walnut it is scanty and sometimes almost non-existent.

According to Mitchell (1978), over 500 species of trees can easily be encountered by anyone looking in parks and gardens in Britain. If special collections in botanic gardens are included, the number rises to about 1,700. The choice of the 33 genera and some 60 species included here has therefore, inevitably, been arbitrary. Because of the current interest and emphasis on conservation, almost all native species which grow to reasonable sized trees have been included, even though some such as rowan and the wild service tree will never become species of major importance. Among exotics, all the commonly planted trees are included, as well as some such as *Nothofagus* and *Pinus peuce* which might become important in the future.

The relative importance of broadleaves and conifers can be gauged by the consumption of timbers of each in the country, and by the extent of areas planted with different species. It is difficult to obtain completely reliable figures for consumption in the UK, but a summary is given in the Forestry Commission's publication *Forestry facts and figures 1988-89*. These are shown in Table 1. Estimates by FAO and the UK Timber Trade Federation indicate that about 82 per cent of all timber consumed is coniferous of which, as Table 1 makes clear, the great majority is imported. In terms of forest areas existing in 1987, 66 per cent was coniferous and 34 per cent broadleaved, amounting to 2.27 million hectares in all, or 10 per cent of the land area of Great Britain. Of this, some 171,000 ha of predominantly broadleaved woodland was classed as "unproductive", and a further 40,000 ha as coppice and coppice with standards.

The principal species, which together made up 80 per cent of the forest area in 1980 (Locke, 1987), are shown in Table 2. From this it can be seen that almost 24 per cent of the forest area of Great Britain is occupied by Sitka spruce and almost exactly the same area by the broadleaved oaks, birches, ash, beech and sycamore. Nine other conifers, of which Scots pine is the only indigenous species, make up a further 33 per cent.

Table 1. Timber consumption in the UK in 1990/91, in millions of cubic metres (under bark) wood raw material equivalent. (From *Forestry facts and figures 1990-91*, published by the Forestry Commission.)

Category	Quantity	% of total
Imported round and sawn wood	20.3	37
Imported pulp	8.6	16
Imported panel products	5.7	10
Imported paper	13.8	25
TOTAL IMPORTS	48.4	88
Home production	6.4	12
APPARENT CONSUMPTION*	50.2	(92)

* Apparent consumption equals imports plus UK roundwood production, minus exports; hence 8 per cent of imports and some home produced timber (4.6 million cubic metres) are (re)exported.

Table 2. Principal species and areas occupied by them in 1980 (devised from Locke, 1987).

Species	Area (ha)	% of total forest area
Sitka spruce	525,901	23.6
Scots pine	241,037	10.8
Oaks	190,044	8.5
Birches	131,391	5.9
Lodgepole pine	127,068	5.7
Norway spruce	116,847	5.2
Japanese/ hybrid larch	111,349	5.0
Ash	79,204	3.6
Beech	75,006	3.4
Sycamore	54,291	2.4
Douglas fir	47,399	2.1
Corsican pine	47,251	2.1
European larch	40,414	1.8
TOTAL	1,787,202	80.1

The remainder of the text of this book provides information about the origin and introduction (where applicable) of each species, climatic and soil requirements, other silvicultural characteristics, provenance, seed production, nursery treatment, yield and timber characteristics. Information about the ages at which trees first produce seeds should be treated with prudence. It is reasonably accurate for trees in even aged plantations or groups at fairly close spacings. Widely-spaced trees often produce seeds earlier than the times quoted.

ABIES ALBA Miller **European silver fir**

This species is very little planted in Britain because it suffers so badly from an aphid, *Dreyfusia nüsslini*. It was important at one time, and may have potential in the future if resistant provenances can be found.

Origin and introduction

Abies alba occurs naturally scattered through the mountains of central, southern and eastern Europe, from the Pyrenees to the Balkans and Normandy. It is especially prominent in the Vosges, Jura, Black Forest and northern Bavaria. It was introduced to Britain in about 1603.

Climatic requirements

High humidity and not too high a temperature are required. The tree does best in Scotland but large specimens can be found almost everywhere (Macdonald *et al.*, 1957). *A. alba* is sensitive to late spring frosts and ideally needs overhead cover when young. Frost-prone flat areas and hollows should be avoided. It tolerates exposure quite well once established. The tree will not withstand atmospheric pollution.

Site requirements

A. alba grows well on deep and heavy soils, but peats and very dry, infertile sands and gravels are best avoided. It does particularly well on calcareous soils which do not dry out, in the west and south-west at moderate elevations. The relationship between site and severity of attack by *Dreyfusia nüsslini* is not fully understood but stands on good soils in cool, moist climates recover better from attack than do those on poorer soils and in warm, dry areas (Varty, 1956).

Other silvicultural characteristics

The tree is very shade-bearing, and like most such species, it is susceptible to damage from spring frosts. It self-prunes slowly and is inclined to produce heavy branches unless they are discouraged by close planting. Early growth, like that of many silver firs, is slow, but it is a considerable volume producer once established. *A. alba* is much less prone to attack by *Heterobasidion annosum* than many conifers. It is said to be a much deeper rooting tree than Norway spruce.

Flowering, seed production and nursery conditions

The tree flowers from May to mid June; seeds ripen between mid September and mid October. The earliest age at which the tree bears seeds is between 25 and 30; maximum production is between 40 and 60 years. There are commonly two or three years between good seed crops. There are about 22,500 seeds/kg (range 17,400 - 41,000). If seed is required for sowing in a nursery, it should be collected immediately it is ripe, before the cones disintegrate. It is normal to stratify, or pre-chill the seed for six or more weeks, prior to sowing in March. Unstratified seed should be sown in January or February (Aldhous, 1972).

Provenance
The tallest sources after ten years growth come from such divergent places as Calabria (Italy), Czechoslovakia and the Swiss Jura (Lines, 1979c).

Timber
The timber is similar to that of Norway spruce and, in continental Europe, is used for the same purposes. It is also known by the same name, whitewood. The average density of the wood at 15 per cent moisture content is about 480 kg/m^3.

ABIES GRANDIS (Lamb.) Lindley — Grand fir

Grand fir is a species which thrives on reasonably fertile, sheltered sites, where it can be one of the most productive trees grown in the UK. Because of its poor timber properties, other species are usually favoured on the sites where it will grow, and it is therefore likely to remain a species of minor importance.

Origin and introduction
Abies grandis is native to Oregon, Washington, north-west California and the coast of southern British Columbia, including Vancouver Island. Separate populations exist further east, in west Montana and north and west Idaho (Steinhoff, 1978). The species achieves its best development in the Olympic Peninsula of Washington. It was introduced to Britain in 1831.

Climatic requirements
In its natural range, the tree does best where conditions are cool and moist. It grows well where rainfall is as low as 750 mm, but responds to higher rainfall. The best stands in Britain occur where precipitation is 1,000 mm or more in the north, and 1,150 mm in the southern half of the country (Macdonald *et al.*, 1957). It is not as demanding as Sitka spruce. Temperatures in Britain are not limiting: it does as well in the north as in the south but the tree is very susceptible to damage by spring frosts, so hollows should be avoided. Grand fir is very sensitive to exposure (unlike *A. procera*) and should not be planted at high elevations; it will not thrive in areas of high atmospheric pollution.

Site requirements
Though requirements are not excessive, *A. grandis* does best on well-drained, moist, light and deep soils of at least moderate fertility. It should be used cautiously on heavy and the infertile soils. It will not thrive on very calcareous

soils, but will grow on dry ones, though on these it may suffer from drought crack. It appears to root much more deeply than many other conifers grown (Carey and Barry, 1975).

Other silvicultural characteristics

A. grandis usually has a very good, straight stem though it often tapers markedly. Older trees suffer crown damage on exposed sites and the species is moderately susceptible to windthrow and windbreak.

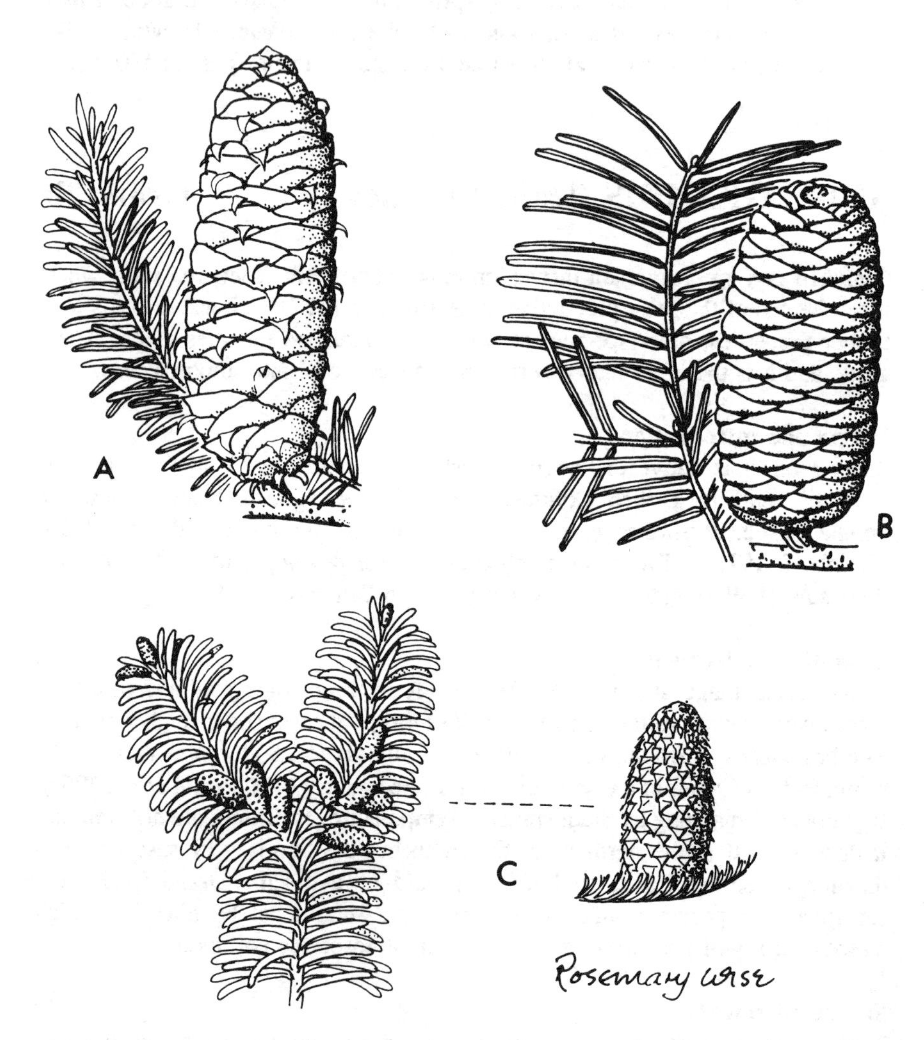

Figure 1. A) European silver fir, *Abies alba*; B) Grand fir, *Abies grandis*; C) Noble fir, *Abies procera*.

It is more resistant to *Heterobasidion annosum* than most conifers but prone to damage and associated wood deterioration linked with the complex of drought, exposure and attack by *Adelges piceae* (Aldhous and Low, 1974), so it is probably safest in the western half of Britain where there is less risk of drought and frost. It is a strongly shade tolerant tree and consequently useful for underplanting. The litter, like that of Douglas fir, breaks down quickly. In common with all other silver firs, early growth is slow.

Flowering, seed production and nursery conditions
The tree flowers in May, seeds ripen in August and September, and should be collected immediately, before the disintegration of the cones. *A. grandis* is rather a poor seed producer in the UK. The earliest age at which it might bear a good seed crop is between 40 and 45. There are commonly three to five years between good seed years. There are about 40,600 seeds/kg (range 26,200 - 63,500), of which 20,000 are normally viable. The treatment of the seed and sowing times in nurseries are the same as for *Abies alba*.

Provenance
In Britain there appears to be a latitudinal cline of increasing vigour in a southerly direction among provenances from the coasts of Canada and the USA. Inland origins grow too slowly to be of much use in Britain (Lines, 1978a). Limited Forestry Commission experiments have shown differences of ten yield classes between the most and least vigorous provenances. Many existing stands are thought to originate from sources at the poorer end of the range (Lines, 1978b). The best growth after six years has been found in provenances from the Olympic Peninsula in Washington (Lines, 1986, 1987).

Yield and timber
On suitable sites, grand fir can be the most productive tree in Britain. Yield classes in excess of 30 have been recorded, and on many sites it will out-produce most other conifers. Unfortunately there are few existing markets for its timber which is much inferior to that of *A. alba*. It is therefore not very widely planted. The average density of the wood is about 450 kg/m^3 at 15 per cent moisture content.

ABIES PROCERA Rheder — Noble fir

Noble fir is a species of great ornamental value, but apart from this its main use in British forestry is on dry, exposed sites at high elevations where Sitka spruce will not grow.

Origin and introduction

The natural distribution is more restricted than those of many other north-western American species, being confined to the Pacific coast from northern Washington to northern California, mostly at elevations between 1,000 and 1,500 m (Franklin *et al.*, 1978). It was introduced to Britain in 1831.

Climatic requirements

In its natural range, noble fir does best where there is a cool, short growing season with abundant precipitation, mostly as snow. There are no well defined climatic limits in Britain but it does best in the north and west, where temperatures and evaporation rates are rather low.

The tree has a very high tolerance of exposure without becoming deformed. This valuable characteristic gives it a potentially useful place in the uplands. *Abies procera* is less moisture-demanding than Sitka spruce which makes it useful for dry, exposed sites. It is less frost-tender than other silver firs, but can be damaged by late spring frosts. Like most silver firs, it will not tolerate atmospheric pollution.

Site requirements

In its native habitat, noble fir is a pioneer species which regenerates after major disturbances such as fire. It does not need high fertility. It grows quite well on boulder tills and some peats, but not in the presence of heather. It tolerates acid soils and does best on well-drained deep, moist soils. Very dry sites should be avoided as it is susceptible to drought crack.

Other silvicultural characteristics

Early growth is always slow, necessitating prolonged weeding and making the species vulnerable to damage from browsing animals. It is moderately windfirm. The tree is relatively resistant to *Heterobasidion annosum*. It can be used for underplanting, but it is not so tolerant of shade as grand fir or hemlock and needs freeing quite early. The main use of *A. procera* in British forestry is on dry, exposed sites where Sitka spruce will not thrive. There is probably no case for large scale planting because of slow early growth.

A. procera is a very ornamental species which is popular as a Christmas tree. It is grown profitably in parts of Europe specifically for its foliage and varieties with blue needles are preferred for this.

Flowering, seed production and nursery conditions

The tree normally flowers in June, seed production is better than the other species of *Abies* in the UK. Seeds ripen from mid to late September, and are dispersed early in October. Cones should be collected in late August or early September because after this the scales loosen prior to the disintegration of the cone, and all the seed can be lost. The earliest age at which the tree bears a good seed crop is between 30 to 35 years, but the best ones are usually between the ages of 40 and

60 years. There are commonly two to four years between good seed crops. There are about 29,800 seeds/kg (range 20,300 - 42,100), of which only 12,000 are normally viable. The treatment of the seed and sowing times in nurseries are the same as for *Abies alba*.

Timber

The timber has a bad reputation, with poor strength properties and a low density (about 420 kg/m^3 at 15 per cent moisture content). It is not naturally durable, but is said to be suitable for general interior joinery and light structural work (Patterson, 1988). If the density of its wood could be increased, through selection and breeding, it could become a much more useful tree.

Provenance

Little is known, but the most promising source appears to be Larch Mountain, Oregon at 975 m, which lies east of Portland (Lines, 1987).

ACER CAMPESTRE L. — Field maple

Acer campestre is the only maple native to Britain. It is found at low elevations in the south, most characteristically in woodlands, hedges and scrub on calcareous soils, especially when heavy, though it grows on a range of other soil types as well. It is frequently associated with ash, together with a ground flora of *Mercuralis perennis*, bluebell, *Sanicula europea*, *Asperula odorata* and similar woodland herbs.

Field maple coppices very well, and coppice conditions favour it because of its tolerance to shade. As a tree, it seldom exceeds about 15 m tall. It is a later seral species, and is never found invading grassland or the early stages of plantations. It is not damaged by frost. The tree flowers in the spring; seed (about 15,000/kg) needs 16-18 months stratification if it is to be grown in the nursery; otherwise it should be treated in a similar way to sycamore. Its wood is slightly heavier than that of sycamore, at about 690 kg/m^3.

ACER PLATANOIDES L. — Norway maple

Origin and introduction

Norway maple is native to central and northern Europe, Asia Minor, Caucasus and northern Iran. It was introduced to Britain before 1683.

Climatic requirements

Norway maple is much less tolerant of exposure than sycamore, and should never be planted on exposed sites. As a result it is much more common, and suitable for planting, in the south of England. It can withstand pollution by smoke and salt spray near the sea.

Site requirements

Its site requirements are very similar to those of sycamore, though it may be less demanding. It grows best on moist, freely draining, calcareous soils. The species is regarded as one of the best broadleaves for planting over chalk, and is said to grow better than sycamore on more acid soils.

Other silvicultural characteristics

It does well in mixture with beech, Norway spruce and western red cedar, but should be planted in groups rather than in intimate mixtures because its rapid early growth may cause suppression of the other species. Norway maple is very sensitive to competition from grass when young and must be kept carefully weeded if it is to succeed.

Figure 2. A) Sycamore, *Acer pseudoplatanus*; B) Field maple, *Acer campestre*; C) Norway maple, *Acer platanoides*.

It grows very fast when young, usually faster than sycamore for the first 30 years or so, and needs to be heavily thinned to maintain growth, but does not reach such large dimensions as sycamore, or live for as long. It is usually described as fairly light demanding.

Like sycamore, it is very susceptible to bark stripping by grey squirrels throughout the pole stage. On suitable sites, and with proper thinning and care, trees of 40 cm diameter at breast height may be achieved in 40 years.

Norway maple regenerates naturally on many sites, but especially on sandy soils where it is rapidly becoming naturalised.

The tree is usually regarded as being more attractive than sycamore: it can have spectacular autumn colour. It is a good pollen and nectar producing tree for bees.

Flowering, seed production and nursery conditions

Norway maple flowers in the spring and produces its first fertile seed between 25 and 30 years, and the best crops every two or three years between 40 and 60, though some seed is produced every year. It is usually ready for collection in September or October. There are about 7,500 seeds/kg of which about 80 per cent normally germinate. For nursery production, the seed should be treated in the same way as sycamore.

Timber

The wood is very similar to that of sycamore, being white and fine textured, though harder. Wavy grained maple is in great demand, and is sometimes used for making musical instruments. The average density at 15 per cent moisture content is slightly greater than that of sycamore, at about 660 kg/m^3.

ACER PSEUDOPLATANUS L. — Sycamore

Sycamore is one of the fastest growing broadleaves and produces a potentially very valuable white timber. It is usually regarded as being very similar to ash in most respects, though it is more frost hardy and less site-demanding. Both species occur on similar soils, they can both be relatively fast growing and both are opportunist species which spread into gaps in the canopy.

Origin and introduction

Sycamore is native to high elevations in southern and central Europe; it extends northwards to Paris and east as far as the Caucasus. Its time of introduction to Britain is uncertain, but it was possibly brought by the Romans, or as late as 1550. It is certainly reported to have been rare in the 16th century and has only become properly established within the last 200 years. There are some 540,000 ha of sycamore in Britain, which make up 3 per cent of the total forest area, including 2400 ha of worked coppice (Locke, 1987).

The ecology of sycamore has been described in detail by Jones (1945), and more recently by Taylor (1985). A feature of the tree is that it is one of the few introduced species which has not only become naturalised but is also spreading, especially in the lowlands. It is continuing to invade many different habitats but particularly valley floor ash woods and calcareous beechwoods, though its requirements do not entirely coincide with those of either ash or beech. Because of its invasive tendencies and the fact that it is an exotic, it has become a controversial species.

Climatic requirements

They compete over much of their ecological ranges. Sycamore is much less sensitive to late spring frosts than most important deciduous species and is very tolerant of exposure to wind on suitably moist and fertile sites, and also to salt spray. It grows at higher elevations than any other broadleaved trees except rowan and the birches, and its potential upper limit is determined more by the presence of suitable soil than by climate. It is consequently useful as a windbreak both at high elevations and in coastal areas. Sycamore is most common in the north and west of Britain.

Site requirements

Sycamore does not grow in soils which are either too dry or too wet. Most writers agree that a deep, moist soil, freely drained and of a reasonably high pH is required: between six and seven has been suggested for very young trees (Helliwell, 1965), but this does not imply that it is particularly site-demanding in Britain. Sycamore grows well and regenerates on neutral shaley soils, on many of the heavier calcareous loams, clay with flints over chalk, and even acid sandy soils, provided they are deep, well-drained but retain some water. A feature of the sites where it does best and regenerates naturally is that the decay of organic matter is rapid and nitrification is free. It does not regenerate where the pH of the soil is less than four, on podzols nor on heavy clays which are gleyed close to the surface, nor where the winter water table comes to within 30 cm of the surface and these sites should be avoided.

Other silvicultural characteristics

Sycamore withstands smoke pollution well. However, more mature trees will rapidly die of suppression if their crowns are not kept free, and the tree has poor powers of recovery if suppressed after the age at which height growth slows considerably. Sycamore can therefore be eliminated rapidly from mixed woods if not kept well thinned. Seed production and natural regeneration in moderately shaded woodlands are often prolific and the species has potential as a shade bearer in selection systems, at least when young. Regeneration is frequently associated with sites where dog's mercury thrives and where elder also regenerates. These tend to be somewhat drier than those where ash does best. Young plants are very intolerant of competition from grasses, so if planted in the open, weeding is important. Natural regeneration in open grassland communities is consequently rare. In mixtures, Stevenson (1985) considered that sycamore and European larch are very compatible with each other in terms of growth rates. Rotations of 65 to 70 years are normal. Sycamore trees up to 80 to 100 years old coppice very well.

Grey squirrels seem to prefer sycamore to most other species and can be exceedingly damaging to trees of up to 25 years old, but the species is attacked by rabbits less than many others, nor is it barked seriously by deer. The sycamore aphid, *Drepanosiphum platanoidis*, is often present on the leaves, where it can produce large quantities of honeydew. Sycamore also suffers from "sooty bark disease" caused by the fungus *Cryptostroma corticale*, which has been described by Young (1978). It can cause dieback and occasionally death but serious outbreaks are, apparently, only likely in the Greater London area. The tar spot fungus *Rhytisma acerina* is often seen on sycamore leaves in unpolluted areas. It does no damage to the tree, apart from reducing the photosynthetic area a little.

Sycamore is the only common tree which produces insect-pollinated flowers (though field maple and lime do too). It is an important source of pollen and nectar for bees.

Flowering, seed production and nursery conditions

The tree flowers in mid April, seeds ripen in September and October, and are dispersed between mid September and mid October. The earliest age at which the tree bears good seed crops is between 25 and 30 and the best ones are usually between the ages of 40 and 60. Some seed is produced every year, but it is produced very plentifully every second of third year. There are about 9,400 seeds/kg (range 5,400 - 15,800), of which 80 per cent will normally germinate. The seed is recalcitrant and is best sown immediately after collection or, if stored (for a maximum of 12-16 weeks), it must not be dried below 35 per cent moisture content. It should then be stratified for eight weeks. Aldhous (1972) states that if it is stratified too soon the seed may germinate early, before the nursery beds are fit to be sown.

Conservation

Sycamore has two characteristics that are believed to have detrimental effects on the ground flora of deciduous woodland: its slowly decaying litter and the heavy shade it casts. In a detailed discussion of these characteristics, Taylor (1985) concluded that in long-established woodland in which sycamore is not totally dominant, it does not cause the extinction of species in the ground vegetation. It may even increase diversity by leading to an increase of those species that are shade tolerant in ash woods, and in beech woods sycamore would allow more light to reach the forest floor. It is in no way the ecological menace that some suggest. Sycamore does support a relatively poor invertebrate fauna compared with most native species.

The meagre evidence that exists suggests that sycamore will eventually be replaced by another species rather than by itself. In ash-sycamore woodland, for example, the regeneration of either species is less likely under their own canopies than beneath each others'. The proportion of sycamore regeneration therefore declines as the proportion in the canopy increases. The species will probably, therefore, alternate with each other and it is unlikely that either would dominate.

Taylor (1985) argues that woodland which has been neglected and hence undisturbed for some time will have very little sycamore. Invasion will be a very slow process to a point of equilibrium determined by the woodland type. In disturbed woodland, invasion is often rapid and aggressive, with sycamore initially becoming a dominant species, but then it declines to a point of equilibrium. Attempts to eradicate sycamore, a widely practised policy in nature reserves, actually provide ideal conditions for its expansion. This is clearly a waste of effort unless it can be shown that it is undesirable for a reason other than that it is a threat to native trees and shrub species.

Provenance

European botanists have recognised several ecotypes and varieties, but none is distinguished in Britain.

Timber

The wood of sycamore normally has no well-marked figure or grain. It is hard, strong and can be worked to a very smooth finish. Its average density at 15 per cent moisture content is about 630 kg/m^3. It is widely used for furniture making and joinery, and is also very suitable for flooring. The white timbers of sycamore and ash are among the most easily saleable of all British hardwoods as there is a market for all grades. Venables (1985) considered more of these two species should be grown. "Wavy grained" sycamore which arises occasionally, can fetch extraordinarily high prices and is used for making good violins and other musical instruments, and for veneers: the intensity of waviness increases with age, so the figure is difficult to detect in young trees.

ALNUS Miller **Alder**

Alders seldom grow to very large trees. The main interest in them in British forestry has been for their potential as "nurses", or soil improvers on sites with undeveloped soil, rather than as timber producers in their own right. They also have values for wildlife conservation, especially along the edges of water. In Britain the alders are neglected.

The number of species recognised by different authorities varies between about 14 and 30. The classification favoured at Kew recognises about 17 species, of which five are shrubs or very small trees. The genus is distributed throughout the cooler regions of the northern hemisphere and it also extends over the mountains of Central America to the highlands of Bolivia, Columbia, Peru, Ecuador, Venezuela and Argentina.

Most alders are pioneer species which invade gaps and clearings in forests. They are capable of direct colonisation even on the most undeveloped of soils, such as new road cuttings and mining spoil. They are usually relatively small and short lived and give way in most instances to later successional species. They are intolerant to shade and competition, and will not grow under a canopy. Only on sites suited exclusively to alder will they succeed themselves.

ALNUS CORDATA Des. **Italian alder**

Origin and introduction

Alnus cordata is native to southern Italy and Corsica, and was introduced to Britain in 1820.

Site requirements

Italian alder will tolerate both dry and calcareous sites such as thin soils over chalk, and reclaimed sites, but it does best on deep calcareous soils and least well on very acid soils. Unlike most species of alder, it is less confined to riverside sites and it appears to have a potentially useful place on dry calcareous soils, either as a nurse for beech, ash or maple, or as a cover tree (Wood and Nimmo, 1962). Its value for these purposes has probably been considerably neglected in the past, and as a consequence it has been planted less than it might have been.

Other silvicultural characteristics

It is a potentially very valuable pioneer species for sites which are being planted for the first time. Its most valuable characteristic, which is at variance with most other alders, is its tolerance of comparatively dry sites. Like all alders, it is a strong light demander and withstands exposure and pollution well.

The coppicing ability of this species is so variable that some authors have claimed that it does not coppice at all. It seems to coppice best if the stem is cut at a height of 30 cm, rather than at ground level.

Italian alder is described as a superb landscape tree, with an upright conical shape and glossy green leaves, and distinctive bark, foliage, flowers and fruit.

Flowering and seed production

Italian alder flowers and fruits at an early age. There are 363,000 seeds/kg, of which half will normally germinate. The seed does not store well.

Timber

The species is suitable for planting for firewood or pulpwood production on coppice rotations.

ALNUS GLUTINOSA (L.) Gaertner — Black alder
ALNUS INCANA (L.) Moench — Grey alder

Alnus glutinosa is an indigenous species found in all parts of Great Britain and Ireland. In continental Europe it extends as far east as Siberia and southwards to north Africa. *A. incana* is found over most of central Europe, extending westwards to France. It was introduced to Britain in 1780.

Climatic requirements

Both species are very hardy indeed and have no serious climatic limitations in Great Britain, though they do not grow well at such high elevations as birch and rowan. *A. glutinosa* is found up to 500 m in the Cairngorms. The frequency of natural populations increases toward higher rainfall areas of the west and north: soil moisture probably exercises more control over local and regional distributions than atmospheric humidity. Even though flushing is early, they do not usually suffer from late spring frosts and are moderately resistant to salt spray.

Site requirements

These alders are very undemanding and will grow on all but the most infertile soils. They are only truely at home on sites not subject to drought, owing to the presence of a permanently high water table: by lakes and streams, and on soils with restricted internal drainage, but they also do well on coarse sands and gravels if moisture is adequate. Alders do best where the pH is above 6, and can withstand short periods of flooding outside the growing season. Neither species grows well on acid peats and they grow badly on dry, sandy soils. *A. incana* grows better on slightly drier and more calcareous sites than *A. glutinosa*. Seedlings will only become established on soil surfaces that come within the capillary fringe of the water table in drier regions, where surface layers remain continuously moist for 20 to 30 days in the period April to June.

Other silvicultural characteristics

The tap roots of both species are apparently able to penetrate anaerobic water sources which other species cannot use (McVean, 1953b), so they sometimes appear to be drought resistant. They are difficult to establish on grassy sites unless weed control is very good.

Like all alders, both species fix atmospheric nitrogen in association with bacteria (*Frankia* spp.) in the very large root nodules and *A. glutinosa* is the only native British nitrogen-fixing tree. This is the attribute which makes them natural pioneers, and enables them to grow well when planted on roadside cuttings, industrial waste sites, drier calcareous soils and on bare limestone outcrops. They can contribute significantly to the nitrogen content of litter and the soil, and can consequently benefit the growth of companion tree species (though it must be said, not always as effectively as some pines benefit other trees). Nodule formation proceeds best in the pH range 5.4 - 7.0 (Ferguson and Bond, 1953) but after they have formed, growth is best at a pH of about 5.4. There is also evidence that the fixed nitrogen, subsequently translocated to the leaves of alders, is returned to the litter in nitrate form and inhibits the development of certain potentially pathogenic fungi of many trees, such as *Poria* and *Armillaria*.

Both species coppice well when young, and *A. incana* also produces suckers especially after mature trees have been felled. They are also notable in that they are among very few hardwoods that are not seriously attacked by hares and rabbits. Moderate grazing levels favour the spread of the trees by reducing the shading and smothering effects of tall herbaceous vegetation on seedlings.

The early growth of alders is very fast which makes their use as "nurses" difficult because the trees being nursed can easily be suppressed and killed. The rapid early growth is due to the speedy development of a large area of leaves and the long period in leaf. Neither species is long-lived, especially on poor sites where the life of *A. glutinosa* may only be 20 to 25 years. Alders in general are very prone to a number of diseases which are probably brought on by stress resulting from the difficult sites where they often grow.

Both species are very sensitive to shading, so that the internal regeneration of alder in woods is unknown except on sites unsuited to the growth of any other trees. Older trees do not respond to delayed thinning.

Alder is the principal, or only food of a very large number of insects and if planted along the banks of rivers and lakes, these insects can provide an important source of food for fish. A policy of planting for this purpose is followed in parts of Canada and the USA. Matthews (1987) considered that alders might have a place in the British uplands for diversifying the large areas of coniferous forests. If they are grown for this purpose, they should be planted in separate blocks because early growth rates are much faster than those of most conifers, which alders therefore tend to suppress.

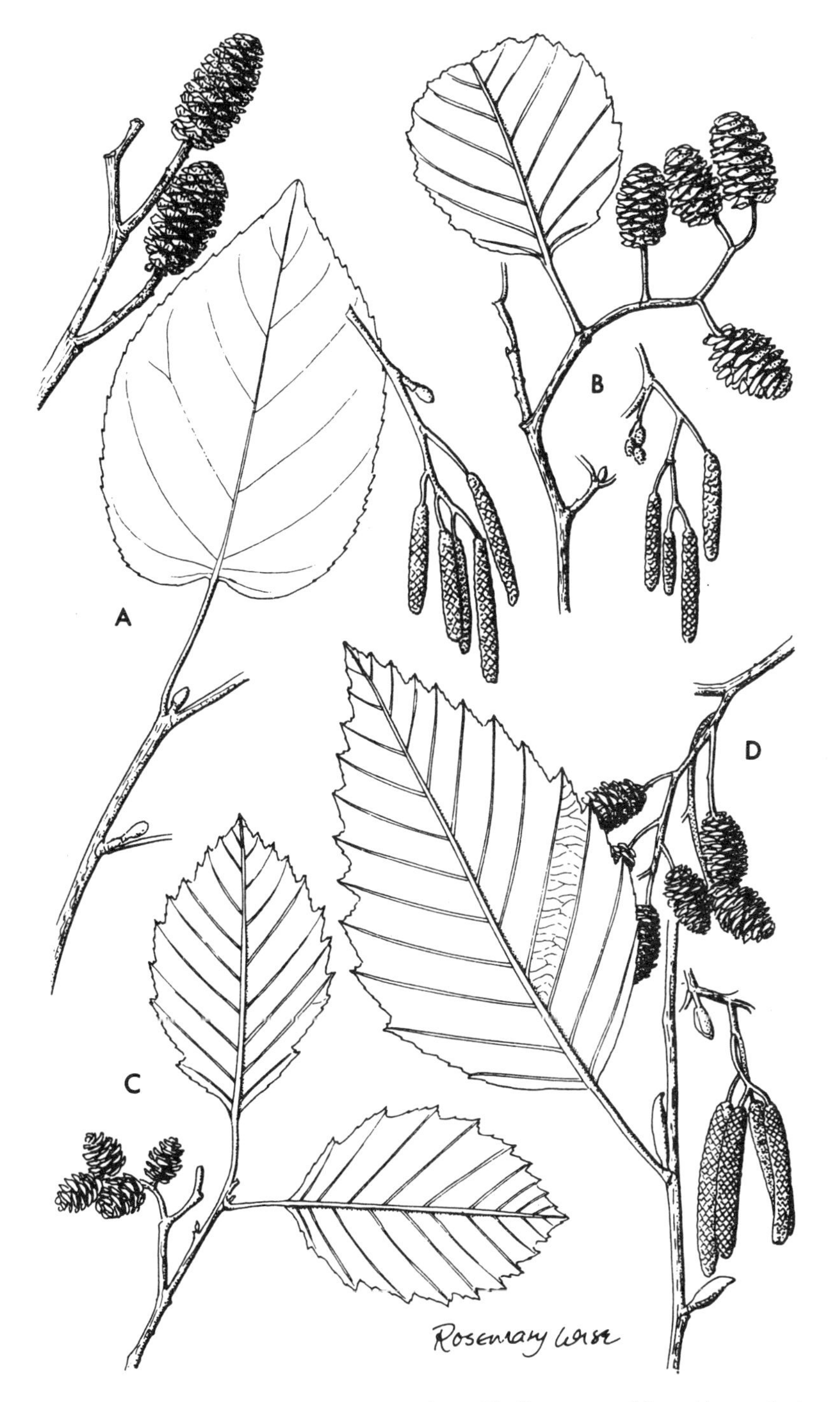

Figure 3. A) Italian alder, *Alnus cordata*; B) Common alder, *Alnus glutinosa*; C) Grey alder, *Alnus incana*; D) Red, or Oregon alder, *Alnus rubra*.

Flowering and seed production and nursery conditions
Both species flower before the leaves are fully out in the spring, the catkins having formed the previous autumn. *A. glutinosa* flowers between early March and late April, and *A. incana* from late February to May. The seeds of both species ripen between September and November, and are best collected for nursery purposes in October. They are dispersed from the time they ripen to early spring. The earliest age at which the trees bear reasonable amounts of seed is about 15 to 20 but the best seed crops are usually after the age of 30. There are commonly two to three years between good seed crops. With *A. glutinosa* there are about 767,000 seeds/kg (range 582,000 - 1,406,000) of which 35 per cent is normally viable, and with *A. incana* the average is 1,460,000 (range 961,000 - 1,980,000), of which only 25 per cent is viable.

Nodule formation on the roots of alder transplants may not occur satisfactorily if they are planted on sites which are being reclaimed, resulting in comparatively poor growth, unless the young plants are inoculated with *Frankia* in the seedbeds of nurseries. An application of extract of crushed nodules, collected from healthy, well grown, tree roots is now recommended to nursery managers as a standard treatment for alder seedbeds (McNeill *et al.*, 1989).

Provenance
McVean (1953a) has described some of the variation which occurs in alder over the British Isles and it is clear that it is considerable. Provenance trials with *A. glutinosa* are now being carried out in many countries and are revealing the variation associated with geographical origin. Matthews (1987) states that the alders are promising subjects for improvement by selection and breeding. Several varieties exist in cultivation.

Area and yield
The fixation of atmospheric nitrogen requires much energy which might otherwise be diverted into vegetative growth: for this reason, alders are not very productive trees, especially on sites where soil nitrogen is low. Yield classes are unlikely to exceed about 10 m^3/ha/year, and as alders are such relative short lived trees, they seldom grow to very large sizes. In some parts of the world there is considerable interest in growing alders on short rotations for fuel, but they are not as productive as other species on better sites. Matthews (1987) estimated that there are perhaps 10,000 ha of alder in Britain.

Timber
The timber is light (about 530 kg/m^3 at 15 per cent moisture content), not very strong, soft and resilient in that a blow causes a temporary depression unaccompanied by a permanent indentation. One of its main attributes is its resistant to decay when submerged in water, but not otherwise. It is sometimes used for making sluice gates and other structures along streams and rivers. It is

a traditional timber for general turnery work, is acceptable for hardwood pulp and is used for making medium priced furniture. Where alder timber is available in large quantities, as in the western USA, it tends to be much better accepted and more widely used. At one time charcoal made from alder was used in the manufacture of gunpowder. Alder is similar to poplar in grain and texture, and is one of the weakest of hardwoods.

ALNUS RUBRA Bong. — Red, or Oregon alder

Alnus rubra is a species which causes excitement at intervals, due to its very rapid early growth which, unfortunately, is extremely seldom maintained. Its most likely value in the UK is as a parent of hybrids, in breeding programmes.

Origin

Red alder has an extensive natural range along the Pacific coast of north America, where it is closely associated with Sitka spruce. It was introduced to Britain in the second half of the nineteenth century.

Site requirements

Site requirements for the spectacular, sustained and rapid growth which occurs so rarely are not really understood. Red alder is reported to have grown particularly well on a clay with flint soil over exposed chalk in Hampshire, and on calcareous boulder clays in Northamptonshire (Macdonald *et al.*, 1957). Like *A. glutinosa* and *A. incana*, it has the reputation of being drought resistant once tap roots have developed. Fraser (1966), for example, has shown one sample of red alder rooting to 89 cm on a surface water gley soil, where Sitka spruce only rooted to 45 cm.

Silviculture

Red alder has proved to be a most disappointing species as a timber producing tree, though growth for the first 10 to 15 years is often very impressive. It can reach 15 m tall in 15 years, after which growth usually declines rapidly and the trees die back, very seldom fulfilling the expectations of early potential. There is little indication that the poor performance is related to unsuitable provenances, as several, at least from British Columbia, have been tried.

A. rubra is more susceptible to spring frosts than either grey or common alder; this may be a contributory factor to the dieback from which it suffers.

Timber

The wood of red alder is similar to that of common and grey alders. It is an important pulpwood in north west America.

BETULA PENDULA Roth — Silver birch
BETULA PUBESCENS Ehrh. — Downy birch

Origin

There are about 60 species of birches in the northern hemisphere, four of which occur in Europe. *Betula pendula* and *B. pubescens* may only be distinct at a subspecific level (Mabberley, 1990), and are native to the whole of Europe, including the British Isles, and to parts of Asia. They reach the northern limits of tree growth. Though not very often planted, natural regeneration is plentiful and they are among the most common of all British forest trees.

Climatic requirements

Both species are extremely hardy and have no major climatic limitations, though *B. pendula* tolerates a drier atmosphere. They both thrive at elevations at which no other broadleaved species except rowan will grow, though the form of the trees can be very adversely affected even by quite low levels of exposure.

Site requirements

The ranges of the two species overlap and they hybridise, but *B. pendula* most commonly occurs on lighter acid soils, heaths, gravels and shallow peats in the drier south and east of the country at low altitudes. It will also grow on heavy clays, and on chalk and limestone soils, but on these it is much less common. *B. pubescens* replaces it on badly drained heathlands and on damper soils generally, especially in waterlogged and peaty conditions in the north and west at higher elevations. On sites where both species flourish, *B. pendula* is considered the faster growing and higher yielding tree. The absence of birch from a site is said often to indicate a deficiency of phosphate in the soil.

Other silvicultural characteristics

Both species are attractive but rather short-lived pioneers. They grow fast when young, but never achieve large dimensions in the UK; trees rarely attain 30 cm at breast height. In parts of continental Europe they will grow much bigger. Prolific natural regeneration can be a nuisance in plantations. Because of their hardiness, birches can be valuable nurses for oak, beech and frost-tender conifers on sites liable to early frosts. They are reputed to be soil improvers (Miles, 1980). Though light-demanding, the birches will grow as an understorey in open forest. They coppice weakly.

The main silvicultural value of birches is their soil improving and nursing qualities, and their considerable amenity value. They have uses in the afforestation of industrial waste sites and for the shelter they can provide for wild and domestic animals, especially in hilly areas. Where there is a market for birch, encouraging natural regeneration can be quite a profitable silvicultural option (using the conventional net discounted revenue criterion) because regeneration is free, rotations very short, and yields reasonable for broadleaved species. Because of their tolerance to a wide range of soils, relatively small sizes, and attractiveness, the birches are suitable trees for planting in small gardens and in urban areas.

They have few pests. They are sometimes browsed by deer, and when a choice of both species is available, *B. pubescens* seems to be preferred. "Witches' brooms" are quite common on birch trees, and are caused by the fungus *Taphrina betulina*, and the species are among the most susceptible of broadleaves to honey fungus.

Figure 4. A) Downy birch, *Betula pubescens*; B) Silver birch, *Betula pendula*.

Flowering, seed production and nursery conditions
The birches flower before the leaves are fully out: *B. pendula* may begin in late March, but the main flowering period is April. Seeds ripen in July and August in *B. pendula* and in August/September in *B. pubescens*. They begin dispersal at once, and seed continues to be released throughout most of the winter. The first good seed crops appear at about 15 years, and the best ones between 20 and 30. Seed is commonly produced in profusion every year, or every other year. There are about 1,900,000 seeds/kg of *B. pendula* (range 1,200,000 - 2,900,000), and 3,570,000/kg of *B. pubescens* (range 1,650,000 - 9,900,000), of which 40 per cent normally germinates. For nursery purposes the seed should be collected immediately before natural dispersal, and stratified for two or three months before sowing in March or early April. Aldhous (1972) states that birches are very sensitive to seedbed surface conditions which should be smooth, well-firmed, with an even, light covering and kept moist.

Provenance
In view of their widespread occurrence, there is scope for breeding to improve the qualities of these species, and work is in progress (Kennedy, 1985). Meanwhile, Lines (1987) recommends that seed should be collected from British stands of good appearance taken from similar or slightly more southerly latitudes than the planting. There are a number of ornamental forms of silver birch.

Area, yield and timber
Birches occupy some 131,000 ha, or 6 per cent of the total forest area of the country, just under half of which is classed as "scrub" in the Forestry Commission census of 1979-1982 (Locke, 1987). The birches are the fourth most common trees in Britain (after Sitka spruce, Scots pine and the oaks).

Maximum yields of about 7 m^3/ha/year may be obtained on the best sites. Birches never grow very large in Britain but the wood can be used for furniture, plywood and veneers (as it is in Scandinavia), and also for pulp and particle board. The wood of birch is fine-textured and of uniform, sometimes decorative, appearance. Fast growth does not reduce wood quality. By most measures it is one of the strongest timbers commonly grown in Britain. It works easily, but is not naturally durable (Lorrain-Smith and Worrell, 1991). Its average density at 15 per cent moisture content is about 670 kg/m^3. There is a limited market for birch wood suitable for turnery, but trees of good form have a potentially much wider range of uses, including plywood, particle board, furniture and high class joinery. If treated with preservatives, it can be used for pallets and fence posts.

CARPINUS BETULUS L. Hornbeam

Origin

There are 35 north temperate species of *Carpinus*. Hornbeam is one of two native European species, extending from the Pyrenees to southern Sweden and eastwards to Iran. In Britain, it is native mainly to the south-east of England.

Figure 5. Hornbeam, *Carpinus betulus*.

Climatic requirements
These are not clearly known, but within the area it is grown, hornbeam is very hardy, even in frost hollows.

Site requirements
The tree is adapted to a wide range of soils, from wet, heavy clays to light, dry sands. It does best on moderately fertile damp sites: Evelyn (1678) suggested it was suited to cold hills, stiff ground and the most exposed parts of woods.

Other silvicultural characteristics
Hornbeam is strongly shade-bearing. Its main silvicultural value is as an understorey tree on moist, and especially wet soils prone to waterlogging. In continental Europe it is commonly encouraged as an understorey to oak to help in the suppression of epicormic branches. Hornbeam coppices strongly and was once important as a coppice species, when it was grown for fuel and basket making. There are still 3,400 ha of coppice being worked in England (Locke, 1987) and it is common in abandoned coppices. Hornbeam, like beech, is less browsed by deer than oak.

Flowering, seed production and nursery conditions
The tree flowers in March as the leaves come out. Seeds ripen between August and November, and are dispersed from then until the spring. They are best collected for nursery purposes in November. The earliest age at which the tree bears seeds is about 20 to 30 years but the best seed crops are usually at intervals of two to four years between the ages of 40 and 80. There are about 24,200 seeds/kg (range 16,000 - 30,800), of which 45 per cent normally germinates. Seed needs stratifying for about a year before being sown in the nursery.

Timber
The wood is hard, strong, tough and white, and finishes very smoothly. It is used for making piano mechanisms, drum sticks, billiard cues, chopping blocks and for flooring as a satisfactory alternative to maple. Its utilisation is not helped by the fact that the tree is frequently fluted and of poor form. Because the wood is very dense (about 770 kg/m^3 at 15 per cent moisture content), and consequently has a high calorific content, it makes an excellent fuel.

CASTANEA SATIVA Miller — Sweet chestnut

Origin and introduction

There are 12 north temperate species of *Castanea* of which the sweet chestnut is the only European species, being native to southern Europe, western Asia and parts of north Africa. It was probably introduced into Britain by the Romans, and is extensively naturalised in south east England.

Climatic requirements

The tree needs warm summer to do well. Northern Britain is too cold, and parts of the east too dry for it to thrive. It does best in southern England, on the right soils. The tree is rather frost-tender and will not tolerate exposure.

Site requirements

Sweet chestnut grows best on deep, fertile, light soils, especially greensands, with ample, but not excessive moisture. It does moderately well on clays and other stiff soils if the sub-soil drainage is good. It does not require high fertility but does badly, and may die back on very infertile soils, calcareous soils, badly drained sites and heavy clays. According to Rollinson and Evans (1987) the ideal pH is 4 to 4.5.

Other silvicultural characteristics

The species is reasonably shade tolerant and is sometimes planted as a "soil improver" on lighter soils. It grows fast when young and coppices very well. Quite large areas are still actively coppiced, especially in Kent and Sussex where it is the last remaining important coppice species in the country. Most of it is worked on a rotation of 12 to 16 years, which is well short of the age of maximum mean annual increment, but provides for the technical requirements of the hop pole and split fencing markets. In the Forest of Dean it is grown on a small scale for pulp, on 25 to 30 year rotations, and where it is not liable to shake, coppice is commonly stored for timber, receiving its first thinning at age 20 to 22. In parts of continental Europe and especially Spain, chestnut is grown as a crop for its edible fruits.

A potentially serious disease is chestnut blight, caused by the fungus *Endothia parasitica*, which at present is confined to southern Europe.

Area and yield

Sweet chestnut is one of the most productive broadleaved species, with mean yield classes of up to 8 m^3/ha/year in some parts of England (Locke, 1978). There are estimated to be 29,000 ha of sweet chestnut in Britain, of which 19,000 ha are coppice (Locke, 1987). Yields of coppice are similar to those for high forest (Rollinson and Evans, 1987).

Flowering, seed production and nursery conditions
Sweet chestnut flowers later than practically any other tree except the limes, in late June and July. The large seeds ripen in the autumn if the summer is warm and late enough, but good crops are very rare, even in southern England. The earliest age at which the tree bears seeds is 30 to 40 years; good production begins after the age of 50. The seeds are very large, with about 239/kg (range 150 - 330). Storage is the same as for oak.

Timber
The main features of the timber of sweet chestnut are its natural durability, even at small dimensions, and the ease with which the wood can be split. At one time extensive areas were grown on coppice rotations for posts, hop poles, fuel and split fencing (Begley, 1955; Evans, 1982). The timber resembles oak, but it is lighter (about 560 kg/m^3 at 15 per cent moisture content), less strong and more easily worked. The heartwood is very durable and a feature of the timber is that, unlike oak, the sapwood is narrow, rarely exceeding about one cm, or about three annual rings. It is a good carpentry and joinery timber. Ring shake is a frequent defect in trees even as small as 40 cm diameter, especially on drought-prone sites. Spiral grain is also a serious problem in large stems. Sound timber is used for furniture and other similar purposes to oak, but it can be difficult to sell anything but the best timber profitably now that the pitprop market virtually no longer exists. Like oak, the wood also has a slightly corrosive effect on metals and it becomes stained when in contact with iron.

Figure 6. Sweet chestnut, *Castanea sativa.*

CHAMAECYPARIS LAWSONIANA (A. Murray) Parlatore **Lawson's cypress**

Origin and introduction

Lawson's cypress was introduced to Britain in 1854, the original seed being sent to Messrs Lawson, seedsmen of Edinburgh after whom it was named. In North America the species is known as Port Orford cedar. It is native to the Klamath and Siskiyou Mountains of north-west California and south-west Oregon, mainly at elevations from 1,200 to 1,800 m. It extends up to 65 km inland on seaward-facing slopes, but is found mostly within 5 to 24 km of the coast.

Climatic requirements

In its native range, Lawson's cypress does best in climates with high precipitation and humidity. It thrives in the western, moister parts of Britain but grows well in drier parts too. The tree is hardy to spring frosts, except when very young, but is not suitable for use on exposed sites. Unlike many conifers, it withstands atmospheric pollution well, which is one reason why it is such a popular garden tree and hedging species in towns.

Site requirements

The tree is not at all demanding, but does best on deep, fertile soils. It is unsuitable for use on peat or dry heathery ground, and grows slowly on heavy clays.

Other silvicultural characteristics

Lawson's cypress is very shade tolerant. In plantations the form can vary from excellent straight stems to a multi-stemmed candelabrum-like habit. Natural pruning is extremely slow. The species seeds frequently and regenerates quite freely. It is also easy to propagate both from cuttings and seed in the nursery. For these reasons it is probably grown much more commonly than it should be, because on sites where it does well, other species such as *Tsuga heterophylla, Thuja plicata* and *Abies grandis* do better. It is less susceptible to damage by *Heterobasidion annosum* than many conifers.

Lawson's cypress has a remarkable tendency to vary in cultivation, and numerous named cultivars exist: surprisingly, practically all of them have arisen in exotic environments in Europe, rather than its native America. This, and the fact that:

1) it is so easy to propagate, from both seed and cuttings
2) it withstands pollution well, and
3) it has a very dense impenetrable habit,

makes it popular as an urban ornamental tree in gardens, and because of its rapid growth, it is even more often used for hedging. Its foliage is valued by florists.

Flowering, seed production and nursery conditions

Lawson's cypress is usually a very prolific seed producer. It flowers in the spring and seeds ripen in September and October of the same year, and are dispersed up to May. The tree bears seed from an early age, 20 to 25 years, and the best seed crops are produced at intervals of two or three years between ages 40 and 60. There are about 463,000 seeds/kg (range 176,400 - 1,322,800), of which 230,000 are normally viable. Cones should be collected as soon as they change from bright green to yellow, and the tips of the seed wings are visible and a light brown colour, normally in September. The seed should be sown in nurseries between late February and mid March: no special treatment is required (Aldhous, 1972).

Timber

The wood, known in the British timber trade as Port Orford cedar, is light yellow to pale brown with no clear distinction between sapwood and heartwood. It has a fine, even texture with a straight grain and a fragrant smell. It is highly resistant to decay, easy to work to a smooth finish, and is stable in service. In its native America, the wood is used for boat building and joinery. The average density at 15 per cent moisture content is about 500 kg/m^3.

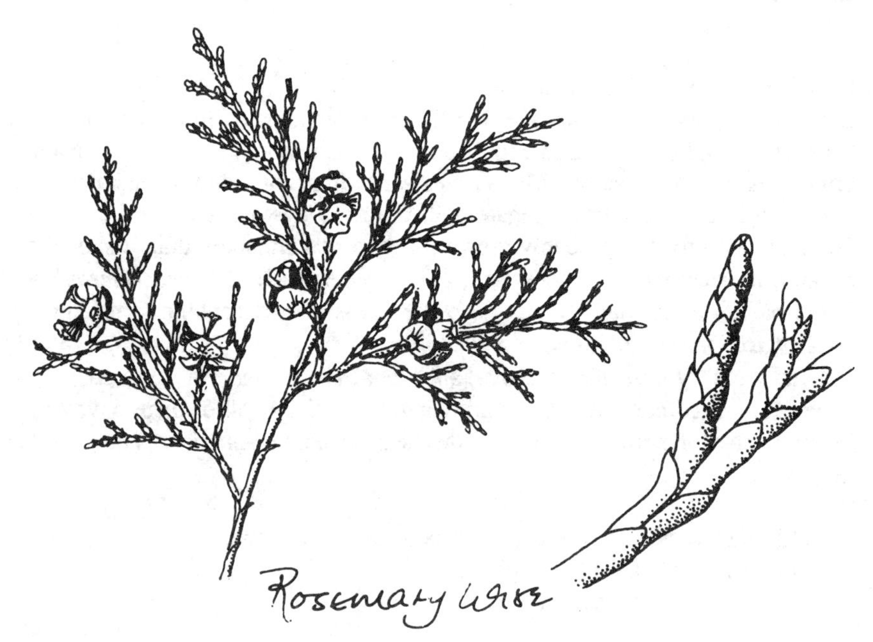

Figure 7. Lawson's cypress, *Chamaecyparis lawsoniana*.

CORYLUS AVELLANA L. Hazel

The genus *Corylus* includes about ten species in the temperate regions of the northern hemisphere. *C. avellana* is native to Europe and common almost throughout the British Isles to elevations of about 600 m; it also occurs in north Africa and west Asia. The hazel is usually seen as a multi-stemmed shrub 1 to 6 m tall, but sometimes as a small tree up to 10 m.

Site requirements
It usually grows on damp, but not waterlogged soils which are no more than moderately acid, but will thrive on dry *basic* soils. Hazel is frost-hardy and reasonably shade bearing, normally being seen as coppice in the understorey of lowland oakwoods, sometimes in ashwoods, and in hedges. It also forms scrub on exposed limestone. It is not particularly long-lived, perhaps 70 to 80 years.

Other silvicultural characteristics
Hazel is particularly susceptible to being browsed by deer and cattle, and must be properly protected after coppicing if the new growth is to survive.

It used to be a plant of considerable economic importance which arises partly from its coppicing ability. It was very widely planted, or propagated by layering. It was then coppiced, both in pure coppice stands, and as the major coppice component in coppice-with-standards systems. As pure managed coppice, about 2,000 stools per hectare were planted, and they were cut on a six to nine year cycle (Hibberd, 1988).

Area and yield
There were estimated to be some 8,600 ha of hazel "scrub" in Great Britain in 1980, and almost 1,500 ha of coppice-with-standards, with hazel as the principal coppice species, and oak by far the predominant standard tree (Locke, 1987). This represents a considerable reduction since the 1947 1949 census when it is believed that there were 12,000 ha of simple coppice and 3,500 ha of coppice-with-standards.

The yields of hazel coppice were investigated in detail by Jeffers (1956). On a five year cutting cycle, with 1,700 stems per hectare, yields are about 15 m^3/ha, and at ten years, about 47 m^3/ha. For well stocked hazel coppice, the production of usable, trimmed, coppice shoots is about 4.5 m^3/ha/year.

Flowering, seed production and nursery conditions
Hazel flowers in the early spring, before the leaves are out. The fruits ripen in September, and fall from October onwards, retaining their viability for about six months. For nursery production, they are usually stratified for three or four months and sown in early April.

Uses

Until the last 100 years, among the many demands for hazel were for its uses in the manufacture of wattles for "wattle and daub" plaster work, sheep hurdles (before wire fencing was common), barrel hoops, garden fencing, thatching spars, fuel for brick kilns and baking ovens, fascines for laying under roads in boggy areas (Forestry Commission, 1956). The older wood was sometimes used by joiners and sieve makers, and the charcoal for gunpowder manufacture. The nuts, which are produced prolifically from about the age of ten years, are edible. In Mesolithic and later human settlements in Europe, it has been suggested that they provided a source of food later taken by cereals (Roach, 1985). Hazel was still grown widely for nut production in Kent at the beginning of the 1900s. The leaves were used for cattle fodder (Schlich, 1891).

Today, hazel has little or no economic value, but as a native component of traditional woodland systems, it is usually considered a species of great nature conservation value. Most of the coppicing which is done today is by naturalists' trusts and similar organisations.

It is an attractive tree, especially in late winter when displaying the yellow catkins.

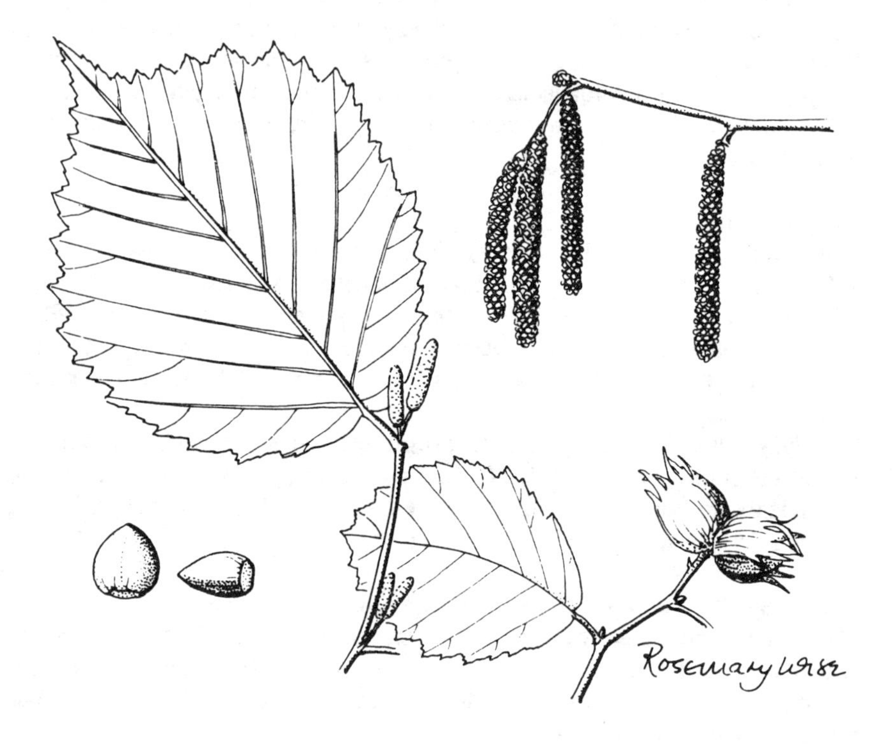

Figure 8. Hazel, *Corylus avellana.*

X CUPRESSOCYPARIS LEYLANDII
(Jackson & Dallimore) Dallimore **Leyland cypress**

Origin and introduction

The tree is an intergeneric hybrid between *Cupressus macrocarpa* and *Chamaecyparis nootkatensis*, which originally arose at Leighton Hall, Welshpool from cones collected in 1888. Several clones now exist, including the golden "Castlewellan" form.

Climatic and site requirements

Although very little work has been done on the silviculture of Leyland cypress, its requirements seem similar in many respects to those of Lawson's cypress. It will thrive on a wide range of soils, including calcareous soils.

Other silvicultural characteristics

Propagation is entirely by cuttings, and plants are therefore relatively expensive. Leyland cypress is light demanding, but tolerant of high levels of pollution and salt spray. It is used widely as a hedging species but has been little used in plantations, consequently little experience with it as a forest tree exists. Limited experience in Bagley Wood, near Oxford, suggests that after thinning plantations of this tree, they become unusually susceptible to windthrow.

EUCALYPTUS L'Héritier **Eucalypts**

There are about 450 species of *Eucalyptus* and almost all occur naturally only in Australia.

As exotics, eucalypts are among the most widely planted and productive of all species both in the tropics and in the warmer parts of temperate regions. Extensive areas are planted for pulpwood and other purposes in southern France and northern Spain. Even at these latitudes (43° to 45°N) they occasionally succumb to exceptionally cold winters, and the stems may be killed, though they quickly regenerate from coppice shoots.

Evans (1986) considered that, in general, the use of eucalypts in forestry north of latitude 45° is likely to be very restricted. The south of England is at 50°N so Britain is probably outside the range where they can safely be grown, but occasional enthusiastic Australian visitors manage to persuade research workers to try new species or provenances at quite regular intervals. The result is that there is a burst of research on eucalypts about every 10 to 15 years in Britain,

Figure 9. Adult leaves and flowers of cider gum, *Eucalyptus gunnii*.

which usually ends in failure after the first severe winter. Species thought most likely to be hardy come from high elevations in the mountains of south-eastern Australia and Tasmania. The current ones being concentrated upon are *E. gunnii, E. archeri, E. niphophila* and *E. debeuzevillei* (Evans, 1986).

Climatic and site requirements

Eucalypts have a reputation for being frost-tender and are damaged or die in cold winters, particularly during sub-zero temperatures which follow a mild spell, or in persistent very cold desiccating winds. *E. niphophila* and *E. debeuzevillei* will often survive midwinter temperatures of -18° to -22°C, but they are susceptible to damage from early autumn frosts. Late spring frosts are less important. Sheppard and Cannell (1987) believe that deaths in winters when temperatures are not too low may be caused by the ground freezing which results in root injury and shoot desiccation.

High fertility is not required. Suitable sites, if any, are most likely to be in sheltered parts of lowland Britain, particularly south-west England, but even there apparently promising trials have been completely killed in severe winters.

Other silvicultural characteristics

Most eucalypts are strongly intolerant trees and so must not be planted in the shade. They coppice very well and benefit considerably from thorough weed control when very young.

FAGUS SYLVATICA L. Beech

Beech is the only large and strongly shade-bearing tree native to Britain: as such, it has an important place in silviculture.

Origin

There are ten species of *Fagus* in temperate parts of the northern hemisphere of which *F. sylvatica* is native to western Europe. Its range extends from southern Scandinavia to central Spain, Corsica, Sicily and Greece, eastwards to western Russia and Crimea, and westwards to Britain where it is now widely naturalised, but believed to be native to south-east England and parts of Wales: Hertford, Gloucester, Powys, Glamorgan, Somerset and Dorset. Brown (1953) has described the performance and silviculture of beech in Britain in great detail.

Climatic requirements

The tree does best in moist, rather mild and sunny climates. Like many shade bearers, it suffers very badly from both early autumn and late spring frosts and can be impossible to establish on exposed, open ground without nurses. It grows well in moderately polluted areas. Once established, it tolerates exposure well, though crowns may become deformed. It has been much used for shelterbelts on hill farms, for example on Dartmoor.

Site requirements

Beech is the characteristically dominant tree on chalk and soft limestone in south-east England, and also on similar terrain across the channel, in Normandy. It does best on calcareous soils if they are not too dry but is by no means confined to them. It is frequently dominant on well-drained loams and sands. Unsuitable sites are either heavy and waterlogged, or infertile, dry and sandy, on both of which it may suffer badly from drought in dry years. It is rare on "neutral" soils (pH 5.0-6.5) according to Peterken (1981). Like many late successional species, young beech competes less successfully than many trees with grasses and other ground vegetation when grown in the open.

Other silvicultural characteristics

Beech benefits strongly from shelter on exposed sites when young. Stem form is often poor, so that planting must be dense to give adequate selection. It is often valued as an understorey to light-demanding species. Beech responds very well to thinning, even at relatively advanced ages, so that when grown in a shelterwood system, the trees left after a seeding cut can put on very valuable increment. It coppices weakly.

Grey squirrels can be very damaging to trees around the age of 40 years, attacking the bases of stems in particular (Mercer, 1984). Stressed trees tend to suffer from beech bark disease, especially on wetter clay soils during and after the pole stage. This disease arises from attacks by a scale insect (*Cryptococcus fagisuga*): later the wounds caused by the feeding activity of the insect may be infected by *Nectria coccinea* (Lonsdale and Wainhouse, 1987). This can kill the bark.

Flowering, seed production and nursery conditions

The flowers, which appear in May, are particularly sensitive to late frosts; seeds ripen in September and October, and fall up to November, often after frost. The earliest age at which the tree bears seeds is 50 to 60, but the best seed crops are usually after 80 years of age and at intervals of as long as 5 to 15 or more years.

Figure 10. Beech, *Fagus sylvatica*.

There are about 4,600 seeds/kg (range 3,400 - 6,400), of which 60 per cent normally germinates. Beech seed has to be stored carefully to ensure it does not heat or "sweat". It can be sown in the autumn if the site is not prone to late spring frosts, otherwise it can be stored by a variety of methods, described by Aldhous (1972), until sown between early and mid March of the following spring. Beech seed cannot be kept longer than this.

Natural regeneration

In continental Europe natural regeneration of beech is common, and is managed under a variety of silvicultural systems. It performs well in the uniform shelterwood system, but because of its tolerance to shade, it can be managed in any of the less regular systems just as well. Beech seed is not dispersed very far from the parent trees, so gaps of 20 m or more are unlikely to be adequately stocked in the middle. The contrast between the apparent ease with which it regenerates on the continent, and the rarity of natural regeneration in Britain is so remarkable that many authors (e.g. Jones, 1952) have attempted to determine the extent to which continental methods might be successful in Britain.

Good mast years, from which natural regeneration arises, usually occur at intervals of ten years or more in Britain, but much more frequently in many parts of continental Europe, including Normandy. It is generally agreed that a good correlation exists between high levels of seed production and the warmth of the preceding summer, provided that the flowers are not killed by a late frost (Matthews, 1963). Most of continental Europe has warm summers and their rarity in Britain is the most fundamental obstacle to obtaining regeneration. Jones (1952) considered that so many British beech woods are even-aged because they arise from exceptionally good and therefore very rare seed years.

A satisfactory seed fall in the Belgium Ardennes is considered by Weissen (1978) to be 200 to 600 seeds/m^2. Enormous losses then occur from fungal attacks (e.g. by *Rhizoctonia solani*), from predation by insects (e.g. *Tortrix grossana*), mice and birds, and from severe winter, or late spring frosts (Engler *et al.*, 1978). In a study near Oxford during the good seed year of 1984, Linnard (1987) found a total fall of 1,316 cupules/m^2 between October and March, indicating a possible fall of 2,632 seeds/m^2. In fact only 1,500 seeds/m^2 reached the ground, and of these half were sound: 8 per cent had been damaged by birds, 4 per cent by small mammals and the rest were empty. By the end of March 1985, only 18 sound seeds remained per square metre, of which six had germinated.

In much of Germany where natural regeneration is practised, the regeneration must be fenced to protect seedlings from roe deer. Apart from damaging the trees, deer browsing encourages herbs and grasses to grow which make conditions much less suitable for regeneration. The young trees must become established within 15 years, which is the average life of a fence.

Though less easily damaged by cold than oak, exposure to temperatures lower than about -6°C can be fatal to beech seed lying on the surface (Jones, 1954). To reduce losses and provide conditions which are suitable for germination and establishment, the ground must be in a receptive condition. The surface should be loose so that the seed is easily buried, and on germination the radicle can readily penetrate the mineral soil. A hard soil surface, or the presence of matted grass, or too much litter all prevent this. Suitable conditions can be achieved by scarifying the soil, or deeper ploughing, or preferably both prior to seedfall (Becker *et al.*, 1978). In Britain, cultivation is likely to be the most important means of making the best use of the light seed crops which normally occur. These treatments protect the seed from frost and *Rhizoctonia* and the roots from frosts and drought in the spring. They also eliminate much of the competing ground vegetation for periods of between one and three years, depending on the species of weeds, by which time the young beech should be competitive enough to be able to look after themselves. On acid soils the addition of phosphate may also help.

The best regeneration in uniform shelterwood systems in France always occurs when a good crop of seed immediately follows a seeding felling: if the felling is done after the seedfall, germination may be just as good, but many seedlings are inevitably destroyed in the subsequent work. The use of shelterwood systems may not be so sensible in Britain, where seed years are so uncertain: the use of group selection systems might be better, but they need a great deal of skill. Above all, British foresters who want natural regeneration must be opportunistic, and be prepared to do the necessary felling when the seed arrives, and this creates obvious difficulties in organising felling, and may flood the market with timber, causing prices to fall.

Successful regeneration is more likely on some sites than others. Jones (1952) considered the best ones to be where there is a moderately shallow loam (about 50 cm deep) over chalk, with a ground vegetation of low herbs such as *Oxalis*, yellow dead-nettle, melic, wood millet and *Holcus mollis*. On the heavier, deeper soils where bramble and coarser herbs are luxuriant, regeneration is more difficult.

Provenance

Very little work has been done on beech provenances in Britain, though Evans (1981) states that Carpathian seed is about average, and the best origins tested come from Soignes forest in Belgium and from the Netherlands. There are several well known ornamental forms of beech, most of which are propagated by cuttings, but copper beeches arise naturally from seed reasonably frequently.

Area and yield

Beech occupies about 75,000 ha or 4 per cent of the forest area in Britain (Locke, 1987). It has mean yield classes ranging from 5 to 6 m^3/ha/year in different parts of the country, and a maximum of 10 (Nicholls, 1981).

Timber and uses

The timber of beech is strong, straight grained, and even textured. The average density of the wood at 15 per cent moisture content is about 720 kg/m^3. It is easily turned and worked and bends well. These properties make it very suitable for the manufacture of furniture, turnery, and formerly kitchen utensils. It is also suitable for flooring, plywood, and constructional work. The beech woods of the Chilterns were originally managed for the furniture industry which was based in High Wycombe and dates from about 1870. In the south of England much of the timber which has grown since the introduction of the grey squirrel in the 1940s is damaged and of little value. In fact, the time of arrival of grey squirrels in any locality can be dated quite accurately from the bases of felled beech trees by counting the annual rings back to the first one which shows damage from gnawing.

Evelyn (1678) commented that beech nuts give a sweet oil "which poor people eat most willingly", and that its leaves "afford the best and easiest mattresses in the world."

FRAXINUS EXCELSIOR L. Ash

Origin

There are 65 species of *Fraxinus* distributed through the north temperate zone. Ash is distributed throughout Europe up to about 64°N, and in north Africa and western Asia. It is native to, and widely distributed throughout most of the British Isles. Clapham *et al.* (1985) consider that it may have been introduced into some northern counties.

Climatic requirements

The whole of lowland Britain falls within its range of temperature and rainfall tolerances, and a combination of exposure and unsuitable soil is probably responsible for setting its altitudinal limit (Wardle, 1961). The tree does best in milder, moister areas: shelter is especially important when young. Ash is frost tender, but because it flushes so late, it often escapes spring frost damage.

Site requirements

Ash usually does best on calcareous loams (pH 7-8) which are moist, deep, well-drained, and have a high content of available nitrogen. Garfitt (1989) states that contrary to popular belief the ideal site is not a damp valley bottom, but a well-drained alkaline soil. This may be a rich deep marl, as in parts of Yorkshire, or a shallow soil over hard fissured limestone, as in the Pennines, or a modest 20 cm of soil over platey limestone brash in the Cotswolds. It is usually absent from sites where the pH of the surface soil is less than 4.4, and less frequent on other acid soils and in drier regions. Really suitable sites tend to be rare and small in area. The species is not suitable for large-scale planting and is probably only worth persisting with on ideal sites where it will grow to large sizes (Stevenson, 1985). Wood sanicle, wild garlic, dog's mercury and elder often indicate appropriate conditions.

Ash is not suited to heathlands nor to the uplands in general, nor should it be planted adjoining arable fields where it is liable to die back severely. Ash seedlings may germinate in profusion on heavy wet soils because they only need a small depth of well-drained soil on which to become established. Periodic water shortages during droughts on these sites lead to poor subsequent growth: they should be avoided because ash never does well on them.

Other silvicultural characteristics

Though ash is strongly shade-bearing for its first seven years or so, it becomes very intolerant later. It does not respond to delayed thinning, and all thinnings should be heavy with the aim of keeping crowns entirely free. The crop should be at its final spacing by age 30 to 35. Pruning is therefore relatively important if the development of large branches is to be prevented. The tree coppices well and natural regeneration is often so prolific that the species becomes invasive.

Ash competes well with *Clematis*, where other trees may become smothered, but not satisfactorily with grasses and so it is not suitable for using on previously unplanted ground.

Ash is often regarded as a useful component in mixtures, though in group, rather than intimate mixtures. It is more rarely successful as a pure plantation species. Cherry is very suitable for mixing with it (Pryor, 1985; Stevenson, 1985), and it also does well in group selection systems with sycamore and beech. In Denmark, scattered ash are grown in beech stands and felled after 70 years, leaving the beech to grow on for another 30-40 years. In Belgium it is often grown with sycamore, cherry, oak, elm, aspen and birch (Thill, 1978).

A bacterial canker (*Pseudomonas savastanoi*) may damage trees badly if they are grown on unsuitable sites, and damage by the ash budmoth may cause forking in young trees. Ash is almost immune from grey squirrel damage, but it is severely browsed by hares, rabbits and deer.

Kent, writing in 1779, said that ash has many enemies "because of the wet, which drips from it, is very noxious to most other plants. And as it shoots it roots horizontally, and pretty near the surface, farmers have a particular dislike to it, because it interrupts the plough." Similarly, Justice (1759) considered that no other tree would thrive under, or near an ash "because it exhausts all the nourishment round it."

Flowering, seed production and nursery conditions

Ash is unusual in that an individual tree may be female, male or hermaphrodite. The tree flowers in April before the leaves are out; the winged "keys" mature in August or September and fall from winter to early spring. The embryos still require several months to develop. For this to occur the seeds must be imbibed with water and maintained under favourable temperature conditions. If seed is collected when green, in July or August, and sown *immediately*, it will germinate the following spring, though rather erratically. It should ideally be collected in October and stratified for 16 - 18 months before sowing in March or April in a neutral soil (Aldhous, 1972). There are about 12,900 seeds/kg (range 8,600 - 16,000), of which 70 per cent will germinate. The earliest age at which the tree bears reasonable seed crops is 25 to 30 years and the best crops are usually produced at three to five yearly intervals between 40 and 60.

Provenance

Practically no work has been done in Britain on selecting high quality trees. Some continental authors distinguish between ecotypes of ash which grow on calcareous sites and those which thrive on damp, clayey sites but such ecotypes have not been identified in Britain. Really superior ash trees with long, clean boles, and excellent crowns are, according to Larsen (1946) and Garfitt (1989), invariably male trees, though work in 1991 in Oxford failed to detect differences in stem straightness between male and female trees. There are some varieties of horticultural interest, such as weeping ash and an entire leaved form.

Area and yield

There are about 79,000 ha of ash in Britain, representing 4 per cent of the total forest area, and of these 1,400 ha are worked coppice. Mean yield classes range from 4 to 6 m^3/ha/year in various parts of the country, with a maximum of about ten. On the best sites, ash is more productive than oak but less than beech. Rotations should ideally be about 60 years, which is when sizes of maximum value are achieved: these are shorter than for sycamore.

Timber

The outstanding property of ash is its toughness after seasoning, as indicated by its bending ability and flexibility. This makes it very suitable for sports goods and the handles of tools such as sledge hammers. The average density of the wood at 15 per cent moisture content is about 710 kg/m^3.

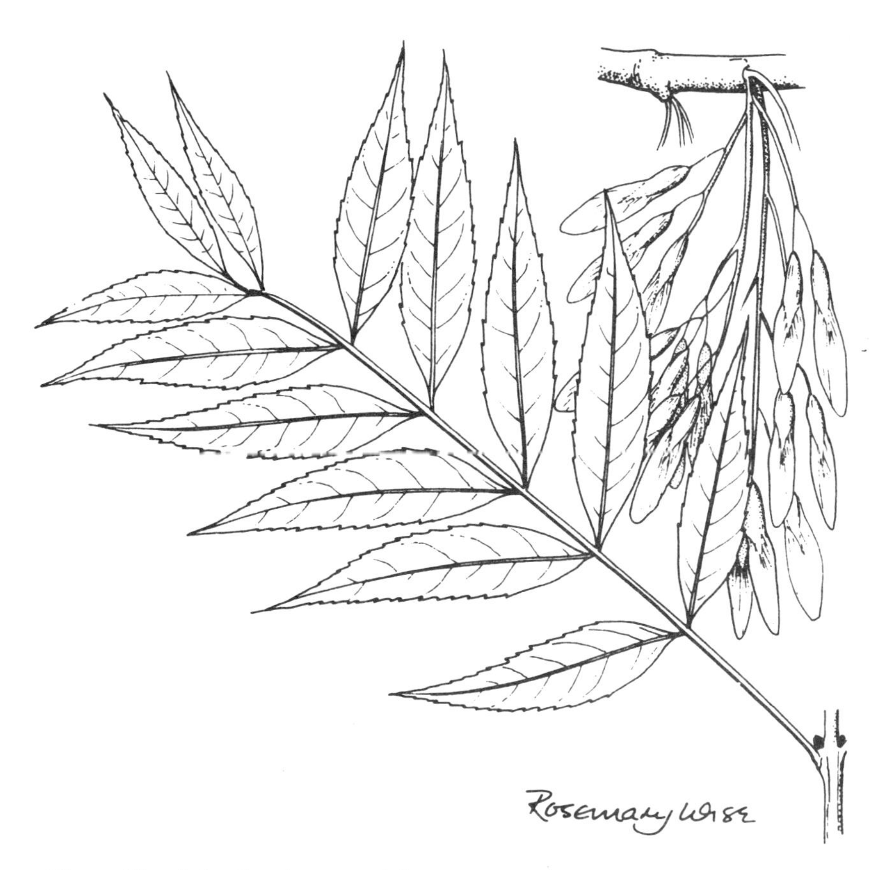

Figure 11. Ash, *Fraxinus excelsior*.

The value of ash timber tends to increase with the rapidity of growth. Timber with 4 to 16 rings/inch is likely to be suitable for most purposes. Faster, or slower-grown trees are unacceptable for the more specialised markets, such as sports goods. White wood is much the most valuable today, but most has some dark coloration in the centre, and this is especially common where the tree is planted on less suitable sites. Evelyn (1678) stated that curiously veined wood could be found. This was prized at one time by skilful cabinet makers. Ash is probably the best firewood, burning well even when green, and is therefore easily saleable for this purpose. Ash, like oak, should not be felled when in leaf otherwise the sapwood is much more prone to attack by wood-boring insects. It is, however, easy to treat with preservatives.

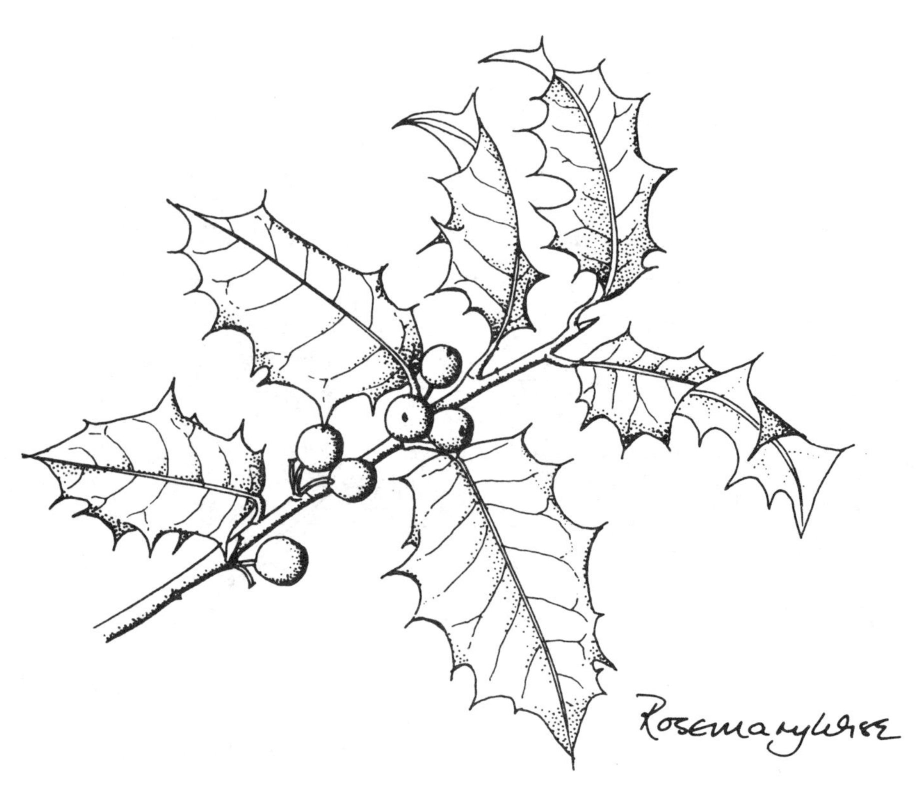

Figure 12. Holly, *Ilex aquifolium*.

ILEX AQUIFOLIUM L. Holly

Origin

There are some 400 species of *Ilex* in tropical and temperate regions. Holly is a common native shrub or small tree (except to Caithness, Orkney and Shetland). It usually occurs in the understorey of beech and oakwoods, where it may often be the dominant species, especially if there is severe grazing. It is also found in hedges, and on rocky hillsides up to over 550 m (Clapham *et al.*, 1985).

Silviculture

Holly is found on a very wide range of soil types but not on wet soils. Marshall (1803) thought that thin-soiled heights seemed to be its natural situation. It is one of the most shade tolerant of all British trees, and withstands pollution well. Holly is very sensitive to transplanting, and can easily be killed by this operation unless it is done with great care preferably in May or September, but anyway during the growing season. Holly is often planted for ornament; numerous cultivars exist with variegated leaves, and different coloured berries.

Flowering and seed production and nursery conditions

The tree is usually dioecious. The flowers are white and appear in summer. The first conspicuously bright red or yellow berries are produced at age 20, but maximum production starts after 40 years: good crops then occur every two to four years. As tradition dictates, they are ready for collection by Christmas when they are widely used for decoration. Seeds never germinate in the first year, and need to be stratified in damp sand for 16 months before sowing. There are about 45,000 seeds/kg of which about 80 per cent normally germinates.

Timber and uses

The wood of holly is much denser than that of any other native hardwood, at about 800 kg/m^3. In 1803, Marshall described it as "being in good esteem among inlayers and turners; it is the whitest of all woods; its colour approaching towards that of ivory." Holly is quite a widely used hedging plant.

JUGLANS REGIA L. **Walnut**

There are 21 species of *Juglans*. *J. regia* is native to south-eastern Europe and spreads to west and central Asia and to China. It is one of the ancient introductions to Britain.

According to Roach (1985), there is evidence that walnuts were grown in Great Britain long before the conventionally accepted time of introduction in the 15th to 16th centuries. Today there are probably fewer trees than at any time since the late 16th or 17th century. Walnut is regarded by some European foresters as a "forgotten" species. There have been campaigns in the distant past for planting it, waged by John Evelyn and others. In the 1700s walnuts are said to have been relatively common in Surrey, Hampshire and other parts of south-east England. Even up to about 1800 planting was pursued with vigour throughout England but by degrees both practice and interest has faded. The reason for this is explained by Marshall (1803), who said that mahogany superseded walnut "in the more elegant kinds of furniture; and beech, being raised at less expense ... and being worked with less trouble, has been found more eligible for the commoner sorts".

Much of the information given below on the silviculture of the species has been obtained from an unpublished report by H.M. Steven, dated 1926, which was prepared at the request of the War Office, presumably reflecting concern about the shortage of walnut for rifle stocks.

Climatic requirements

According to Rebmann (1912) the northern limit for the cultivation of walnut in Germany is between 44° and 52°N. If this is true for Britain its use would be limited to the south of England, though Mitchell and Jobling (1984) state that it can be found everywhere except in the Scottish Highlands, but it is most frequent in south Yorkshire, Lincolnshire, Devon, Somerset and Dorset. It is said to be a very exacting species, requiring both warmth and a long growing season. Though winter cold is not a particular threat, it is liable to suffer badly from late spring and early autumn frosts, and unseasonable frosts in general (Lake, 1913).

Young shoots and flowers are easily damaged by spring frosts of -1°C in Britain. These frequently cause a failure of crops of nuts, hence the reliance on imports. Several varieties (listed by Roach, 1985) are relatively resistant to frosts and most of these originate from the famous French walnut growing region of Isère, near Grenoble.

Site requirements

Walnut is possibly the most site demanding of all potential timber trees in Britain. In 1597, Gerard wrote "The walnut tree groweth in fields neere common high waies in a fat and fruitful ground, and in orchards; it prospereth on high fruitful banks; it love not to grow in waterie places." Evelyn (1678) said it delights in a dry, sound and rich land, especially if it is chalky. Low water tables and good drainage over deep sandy loams are particularly favourable (Batchelor, 1924) and a general view is that walnuts do well where beech thrives. Sites to avoid are light, sandy soils and heavy soils in general (Klemp, 1979), shallow soils over chalk, peaty soils and damp situations generally.

One of the difficulties of growing walnuts in Britain is the scarcity of available and really suitable sites: most potential ones are used for arable agriculture.

Other silvicultural characteristics

Walnuts, like poplars, are very intolerant of competition and, for this reason, are best grown in unthinned orchard conditions, planted at 12 x 12 m spacing, or as isolated trees or hedgerow trees, rather than in more normal forest conditions. If grown in forests, they should be planted in groups and the crowns need completely freeing by the age of 30 to 40 years. Walnut has strong phototropic tendencies and the leading shoots grow towards gaps in the canopy if the trees are shaded.

Figure 13. Walnut, *Juglans regia*.

Like cherry and the poplars, they benefit considerably from careful early weeding. The frost-tender nature of the walnut indicates the desirability of using nurses when young but it is so strongly light-demanding that this is difficult unless the nurses are later shade-bearers. Mixtures with beech and hornbeam were suggested by Schwappach (1926).

Walnuts "bleed" profusely after pruning and it is usually recommended that any necessary pruning is done in the late summer (July and August) or early autumn, before the leaves fall, in order to minimise this. Many species of walnut have the reputation of producing allelopathic chemicals which are washed off their leaves and prevent the growth of potentially competing vegetation. The lemon-scented leaves deter insects. When horses were used for transport, walnuts were quite commonly planted for them to rest under, away from the nuisance of flies.

Walnuts are prone to a bacterial blight, *Pseudomonas juglandis*, when grown in large numbers together.

Spence and Witt (1930) suggested that this species might profitably be planted bordering arable fields with the prospect of crops of nuts during good years and an annual increment in timber value.

Seed and nursery conditions

A characteristic of walnuts is that they produce very large, strong tap roots and, when young at least, few lateral roots. Fibrous roots are slow to develop. This makes them very difficult to transplant successfully. Most authors (e.g. Popov, 1981) state that for these reasons planted seedlings are particularly difficult to establish. Plants from direct sowing of pre-germinated seed are faster-growing and have better root systems. When nursery stock is used, it should be planted out as young as possible. By contrast, Aldhous (1972), while recognizing the care which is needed in lifting two year old plants from seedbeds, recommended they should be moved to transplant beds for a further two years. If stems exceed about 2 cm in diameter at the base, it is often prudent to cut them off at 2 to 3 cm above the ground and allow a new shoot to grow. Some authors also advocate pruning the taproot too (e.g. Macdonald *et al.*, 1957).

There are about 65 - 180 seed/kg, of which 75 per cent will normally germinate if they are stratified immediately after harvesting until the spring, and sown when the seed is on the point of germinating. Bad and small nuts should ideally be rejected. Fertile, near neutral soil is needed for good nursery stock.

Timber and uses

Walnut wood is very stable. It scarcely warps or checks at all, and after proper seasoning shrinks and swells very little. It is easy to work, and holds metal parts with very little wear or risk of splitting. It is uniform, and slightly coarse (silky) in texture, which makes it easily held. It is strong without being too heavy (density at 15 per cent moisture content is about 670 kg/m^3), and can withstand considerable shock. Its colour is dark (so it does not show the dirt) and it often has a most attractive figure. Many of these attributes make it the most valuable wood for gunstocks.

There is considerable interest in France in planting walnuts for their potentially extremely valuable timber, in the belief that decorative temperate woods are likely to be in much greater demand as tropical supplies decline. This echoes what Marshall said in 1803: "were the importation of mahogany to be obstructed, the walnut it is probable would become a very valuable wood". The wood is used for making veneers and high quality furniture. Highly figured veneers are used for cabinet-making and decorative panels. Walnut was one of the finest woods for aeroplane propellers in the early days of aviation. Between 1978 and 1986 walnut logs sold in Germany for between 270 and 3400 DM/m^3 (£90 - £1,100).

The large burrs which are found occasionally on trees are particularly valuable for veneers. A notable source of these, which have flame figures low in the stem, occurs at the junctions of grafts of *J. regia* and some Californian species harvested from over mature (40-60 year old) trees in Californian nut orchards. Such burrs were said to sell for between £5000 and £10,000 each in 1992.

According to Spence and Witt (1930), the great majority of English walnuts suffer from insufficient development of the normal oil content of the nut, and excessive moisture in the kernel, probably caused by an insufficiently long and hot summer, particularly in more northern areas. This makes them more suitable for pickling. For pickling they should be collected while still young and tender, before the nuts become woody and while they can readily be pierced by a needle without the soft shell which is forming inside the husk being felt - usually towards the end of June or early July.

The oil pressed out of walnuts was at one time used in some parts of France instead of butter and olive oil (Reneaume, 1700). The husks of the nuts when boiled produce a dark yellow dye which was once much used for colouring wood, hair and wool. It was also used by fictional heroes for disguise.

LARIX Miller — Larches

There are nine species of larch, all of which occur within the cooler parts of the northern hemisphere.

The attractions of the species used in the UK arise from the ease by which they can be established on weedy sites, their rapid early growth and the general usefulness of their timbers, even at very small sizes. As the only commonly planted deciduous conifers, they also have considerable amenity values in spring and autumn, in particular.

LARIX DECIDUA Miller — European larch

The tree is native to the mountains of central Europe, from the Alps to south Poland, the Carpathians and Croatia. Many races exist but four main groups from the Alpine region, Sudetan mountains, Poland and Tatra are usually recognised. The date of introduction is not known, but the species was in Britain by 1629.

Climatic requirements

European larch does best where the atmosphere is reasonably dry and sunny, but high temperatures are not required. It will grow well at high elevations if the site is not too exposed. Because it often flushes as early as mid March, particularly in milder regions and at low elevations, the tree is very susceptible to frost damage in March and April which, apart from killing foliage, may allow infection by canker. Apart from this danger, European larch has no particular climatic limitations in the UK. However, it will not thrive in conditions of low rainfall, on soils which are not retentive of water, such as lowland heaths (Macdonald *et al.*, 1957). It is only moderately tolerant to pollution, and not to sea winds.

Site requirements

This is an exacting species requiring at least moderate fertility and soils must be moist but freely draining. Sites to avoid include poor and non water-retentive sands, peats, heavily leached soils and soils over chalk, and also areas carrying dense heather. The tree needs to root deeply, so badly drained sites prone to waterlogging must also be avoided. The species does best on middle and higher, but sheltered slopes.

Other silvicultural characteristics

European larch is a very light-demanding, pioneer species which, according to Anderson (1950), has high water requirements after flushing and so needs a continuous supply of soil moisture. It tends to die on drought-prone sites.

Thorough weeding in the year or two immediately after planting is important, but after that, European larch is more competitive than many other species, and weeding can be relaxed. This is one reason why it is a popular plantation species on many estates.

Because it is so strongly light demanding, thinning must begin as early as age 12 to 15 years, and must be heavy. Natural pruning takes place rapidly.

Its relatively light crown and compatible growth rate makes it a useful nurse for oak on sites where it does not grow too fast.

Larch canker, *Trichoscyphella wilkommii*, is a serious and often complex problem on unsuitable sites (Pawsey and Young, 1969) and it is also relatively susceptible to *Heterobasidion annosum* and to the leaf cast fungus, *Meria laricis*. European larch is unlikely ever to become a very important species in Britain because Japanese and hybrid larches perform better in most, but not all, areas where it will grow.

Provenance

The most suitable provenances for use in Britain come from the border regions between Czechoslovakia and Poland, between 300 and 800 m in elevation (Lines and Gordon, 1980). Because of the difficulties of collecting seed in eastern Europe it is still virtually impossible to obtain. Registered seed stands in Britain are therefore the first choice but it is often difficult to obtain adequate supplies from these. Most British stands are of high Alpine origin and are susceptible to canker. Seed should not, therefore, be collected from them.

Flowering, seed production and nursery conditions

The tree flowers in March and April, and flowers are often damaged by frosts. Seeds ripen between September and December, and fall between October and the spring. The earliest age at which the tree bears seed is 20 to 30 years and the best crops are usually at three to five yearly intervals between 40 and 60. There are about 158,700 seeds/kg (range 92,600 - 269,000), of which 60,000 are normally viable. For nursery purposes seed is usually sown in mid to late March. It requires no pre-treatment unless it is dormant, in which case it should be stratified for three weeks (Aldhous, 1972).

Area and yield

There are 40,000 ha of European larch plantations in Britain, representing about 2 per cent of the total forest area (Locke, 1987). Mean yield classes in different regions range from 6 to 9 m^3/ha/year (Nicholls, 1981), with a maximum of 12.

Timber

Details are given under Japanese and hybrid larches. The average density of the wood of European larch at 15 per cent moisture content is about 590 kg/m^3.

LARIX KAEMPFERI (Lamb.) Carrière Japanese larch
LARIX X EUROLEPIS Henry Hybrid larch

Origin and introduction

The natural range of Japanese larch is confined to a small region centred in Nagano Prefecture on the island of Honshu in Japan, between 35° and 37°N, at elevations between 1,200 and 2,400 m. The species was introduced to Britain in 1861.

Hybrid larch is a cross between *L. decidua* and *L. kaempferi*. It first arose about 1895 at Dunkeld, Tayside and was noticed in 1904. Silviculturally, it is usually considered similar to Japanese larch, but in terms of growth it is superior to either parent.

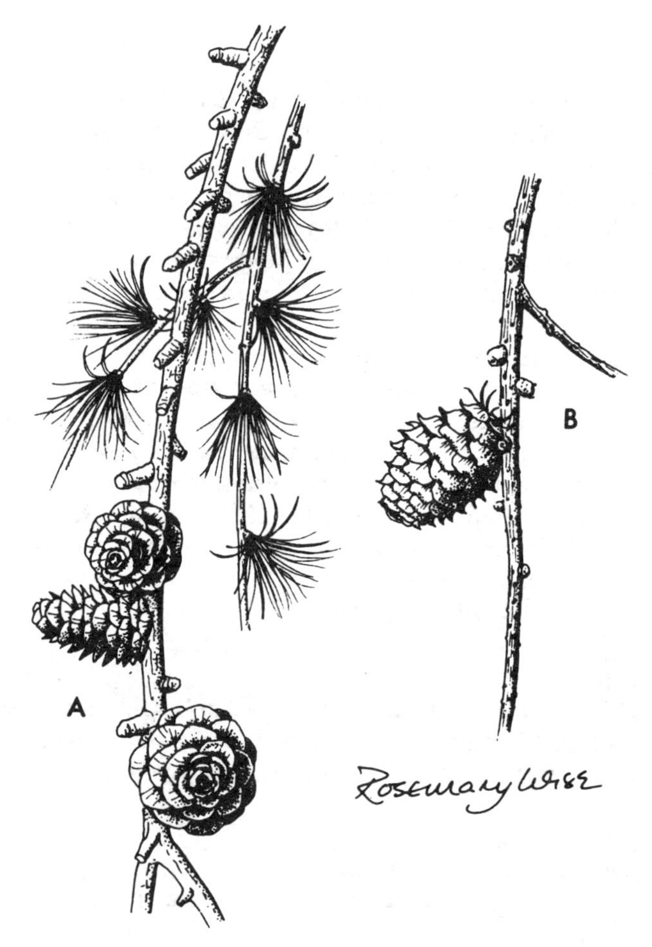

Figure 14. A) European larch, *Larix decidua*, B) Japanese larch, *Larix kaempferi*.

Climatic requirements

In the native range of Japanese larch, there is a summer rainfall climate with over 1,000 mm a year. In Britain, it needs a high atmospheric humidity during the growing season, unlike European larch. It does best in hilly districts in the milder, wetter west coast regions. Areas where the precipitation is substantially below 700 mm should be avoided unless the conditions of soil moisture are particularly favourable. Trees stop growing and young ones may die in periods of severe droughts. The range in Britain is not limited by temperature but these species harden off late in the autumn and are therefore susceptible to early autumn frosts (Macdonald *et al.*, 1957). Japanese and hybrid larches, but especially the latter, tolerate exposure quite well on good soils, and they withstand salt spray remarkably well too. They are also better than European larch at tolerating pollution, though none of the larches is very good in this respect.

Site requirements

In its natural range, Japanese larch is a pioneer species which colonises recent volcanic ash deposits. Japanese and hybrid larches are more accommodating than European larch and they thrive over a wide range of conditions, but do best on well-drained, but moist moderately fertile soils which are not too heavy. They are suitable for upland planting if the soil is reasonably well drained. The presence of dense bracken is often used as an indicator of a suitable site. Dry sites in the south and east of England, and elsewhere should be avoided, and also badly drained land where the trees may become very unstable and, in dry years, suffer from drought. In exposed situations, especially where the soil is fertile, the trees tend to lean and twist, to the extent that they are useless for sawn timber production.

Other silvicultural characteristics

In most respects these species are similar to European larch. Their main attractions are their ease of establishment, rapid early growth, amenity value and the possibility of early financial returns. Their fast early growth makes them particularly useful on old coppice areas and sites where weeds, especially bracken, are a problem. They are valuable for fire belts and sometimes as nurses for more tender species, but if they are used for this purpose, they must be removed early enough. Natural pruning is good. They are strongly light-demanding and consequently need to be thinned well: at least one third of the total stem length should be live crown. Though seed production is good, natural regeneration rarely persists in great quantities.

They are susceptible to *Heterobasidion annosum* and also, particularly, to *Armillaria* spp., however both are generally free of the canker which is so damaging to many European larch provenances. If conditions are not favourable these larches do not go into check like spruces, but tend to die quite quickly. Larches are, on occasions, damaged by squirrels.

Provenance
Most seed of Japanese larch is now collected from selected British stands. If it has to be imported, suitable areas for collection in Japan are given by Lines (1987). Hybrid larch seed is usually very difficult to obtain, though work on producing F1 hybrids from cuttings has proved successful and may soon be commercially viable.

Flowering, seed production and nursery conditions
Japanese and hybrid larches flower in late March and early April; flowers are often damaged by frost. Seeds ripen in September, and are dispersed naturally over the winter. Seed can be collected in September. The earliest age at which they bear seed is 15 to 20 and the best seed crops are commonly at three to five yearly intervals between 40 and 60. There are about 253,500 seeds/kg (range 125,700 - 335,100), of which 100,000 are normally viable in Japanese larch, and only half this number in hybrid larch. The treatment of seeds for nursery purposes is the same as for European larch.

European larch is similar in most respects, but its seeds are ready for collection somewhat later, usually in December.

Area and yield
Larches are not high volume producers. Mean yield classes range from 7 to 11 in different parts of Britain (Nicholls, 1981), with a maximum of little more than about 14 m^3/ha/year. Japanese and hybrid larches cover 111,000 ha (5 per cent of the forest area) and together are the seventh most common species. The disadvantages of the rather low yields are to some extent offset by the very early age at which profitable production can begin, and the short rotations.

Timber
The timber of European, Japanese and hybrid larches is noted for its hardness, natural durability and strength. It tends to distort on drying and is also very resinous. Japanese and hybrid larches are described as being milder than European, with a tendency for the earlywood to crumble and tear during sawing. All species are used for outdoor work, including transmission poles, the exterior of buildings where something more durable than spruce is needed, and for boat building. Even very early thinnings can be sold profitably for rustic work in gardens. Although durable, preservative treatment can be very worthwhile. The average density of the wood of Japanese larch at 15 per cent moisture content ranges between 530-590 kg/m^3, and of hybrid larch, about 480 kg/m^3.

MALUS SYLVESTRIS Miller **Crab apple**

There are about 25 species of *Malus* within north temperate regions. The crab apple is a small light-demanding tree which occasionally grows up to 10 m tall. It is native to Britain and is found throughout England and Wales, but is rare in central and north Scotland. It occurs particularly in oak woods, and in hedges and scrub up to elevations of almost 400 m. It grows on a wide range of soils, from acid to basic, and clays to sands.

In the wild, it often hybridises with cultivated apples of which three species have had a major influence: *M. sylvestris* and *M. pumila* (originally from the Caucasus and Turkestan), and in recent years, *M. baccata* (a native of Russia). The genetic make up of the apples cultivated today is very complex because of the easy hybridisation between species which has occurred naturally, and over thousands of years of selection and breeding by man (Roach, 1985).

It is believed that the druids planted apples in the vicinity of their sacred groves of oak trees, possibly because they served as hosts for mistletoe, which was of great importance to them. The tree flowers in the spring.

Timber

Well grown trees provide timber of excellent quality. It is a dense wood at about 720 kg/m^3 at 15 per cent moisture content. It has a fine uniform texture with no distinct heartwood. It is very resistant to splitting, but tends to distort unless dried very slowly, but when properly seasoned and kept dry it holds its shape well enough to be used for precision work: carving, wood engraving, recorders, tool handles and turnery. At one time it was used for making set-squares and other drawing instruments.

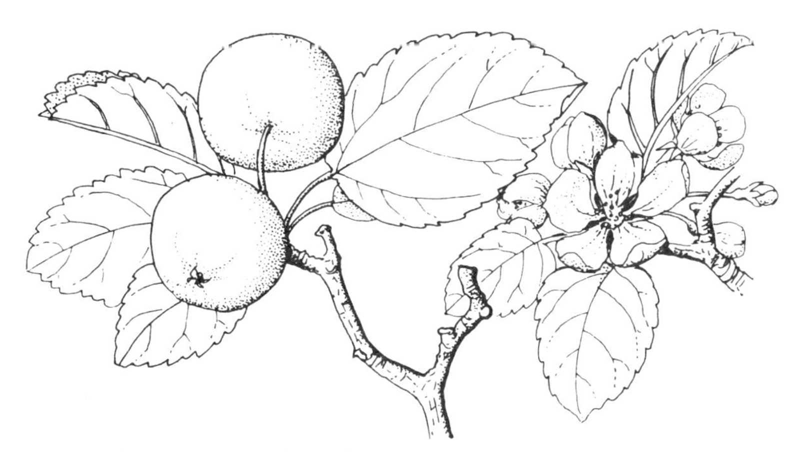

Figure 15. Crab apple, *Malus sylvestris*.

NOTHOFAGUS OBLIQUA (Mirbel) Oersted
NOTHOFAGUS PROCERA (Poeppig and Endl.) Oersted
Chilean, or southern beeches

Nothofagus is a genus of the southern hemisphere where there are 35 species distributed between New Guinea, New Caledonia, New Zealand and the temperate parts of Australia and south America.

No species has an established place in British forestry but two, *N. obliqua* and *N. procera*, are attracting interest because of their potentially high levels of production, even on rather poor sites. Though broadleaved, they can grow as fast as many conifers on some sites and therefore offer the possibility of getting away from the extensive coniferous monocultures to which many people object. However, much more research and an extended period of trials are needed before this can happen.

Origin and introduction

Of the 40 species of *Nothofagus* in the southern hemisphere, only those from Chile and Argentina are likely to be of value in Britain. Trees of the genus *Nothofagus* are the dominant components of the temperate and subantarctic forests in Chile from 33° to 56°S and on the drier side of the Andes, south of approximately 39°S. Ten species occur in this region, one of the most abundant of which is *N. obliqua* which originates from between 33.5° and 41.5°S in Chile, from an area which is climatically very variable because of large altitudinal differences (Donoso, 1979). It was first brought to Britain in 1849 but died out and was reintroduced in 1902. *N. procera* comes from Chile and south-west Argentina and was introduced in 1931.

Climatic requirements

The most serious limitation to the use of *Nothofagus* is their lack of cold hardiness. They can just be grown in Britain, but the species are almost unknown even in the most mild parts of continental Europe, because they fail to survive. Murray *et al.* (1986), found that both species harden very slowly in the autumn and are damaged by temperatures of -14°C in mid winter. Most trees are killed by temperatures of -20°C. The trees deharden very early during the frosty weather of February and March prior to bud burst in the latter half of April. By contrast, the native beech hardens rapidly in September, is undamaged by frosts well below -20°C and does not deharden until late April. Such characteristics almost guarantee that *Nothofagus* will suffer severely from frost damage at least once during a rotation in most parts of Britain. Frost can cause dieback of shoots and consequent multiple leaders, it can kill the cambium and the resulting damage ranges from irregular swellings to large open fissures which allow entry to fungal pathogens; trees are also often killed.

If individuals could be selected that are 3° to 6°C more frost hardy than the population mean, they would avoid frost damage in most lowland regions. Apart from frost damage, both species have done well in areas with rainfall ranging from 700 to well over 1,000 mm. They are sensitive to prolonged dry spells which can cause dieback of the shoots and branches. In Scotland, dieback and death, mainly from cold, is common. The species seem most suited to the lowlands and hills up to 300 m in the south and west of Britain, but not to the most severe upland climates, especially sites exposed to cold north and east winds.

Site requirements

Neither species is too exacting. They grow well on a wide range of soils, from deep sands to heavy clays. Sites to avoid are those with severely impeded drainage, shallow soils over chalk, calcareous clays, and acid peats. On the more difficult sites, *N. obliqua* may succeed where *N. procera* fails and it tends to produce the best results even in the south of England (Lines and Potter, 1985). *Nothofagus* can grow faster than beech on all but calcareous soils. Good growth rates are obtained on sites marginal for oak.

Other silvicultural characteristics

Plants are very susceptible to drying out between lifting in the nursery and planting: the use of anti-transpirants on stems is sometimes recommended. Early and thorough weeding is essential for good establishment, and growth is best where there is a light overstorey, but this must be removed after 2-3 years. Neither species suffers too badly from grey squirrel damage, even in areas where beech is badly attacked. For nature conservation purposes, *Nothofagus* is considered better than beech and most conifers (Wigston, 1980). They coppice well if the shoots are in full light. *N. obliqua* has the better form, being narrower, less heavily branched and it nearly always has a single, good straight stem, if undamaged by frost.

Seed production and nursery conditions

Both *Nothofagus* species produce reasonably good natural regeneration, and they hybridise with each other. For nursery production, Aldhous (1972) states that the seed should be stored dry, and stratified in early March for three weeks before sowing. *N. obliqua* produces about 120,000 seeds/kg and *N. procera*, 100,000. Germination percentages are usually low, at between 25 and 35 per cent.

Provenance

There is still a serious lack of information about the variability of these species. Many of the earlier imports of seed came from areas which are too different from the British climate for the trees to succeed, probably from too far north in Chile. The serious stem cankers and dieback which result from cold winters may possibly be overcome if more southerly provenances are used. Seed from low elevations, south of 38°S is recommended by Tuley (1980). Recommendations for the use of various provenances in Britain are given by Potter (1987).

Yield

N. obliqua and *N. procera* have no equals among broadleaves at their highest levels of volume production and in some situations they compare favourably with the best conifers. The mean yield class in Britain is 14 m^3/ha/year and maximum local yield classes of 20 occur in places, such as the Forest of Dean. Rotations are unlikely to exceed 45 years (Christie *et al.*, 1974). Rates of height growth, though not necessarily of volume production, may be equalled by Japanese larch, and among broadleaved species by cherry, sweet chestnut and poplar.

Timber

The timber of *N. procera* of Chilean origin, known as rauli in the British timber trade, is said to be the best of all *Nothofagus* species, and is comparable to, though lighter than beech (about 470-560 kg/m^3 at 15 per cent moisture content compared with 720 kg/m^3). One of the main uncertainties with these species in Britain is whether the very fast-grown trees will prove to be sufficiently useful to have any significant value. Early indications are that this is unlikely. The wood of British-grown *N. obliqua* has a density of about 600 kg/m^3.

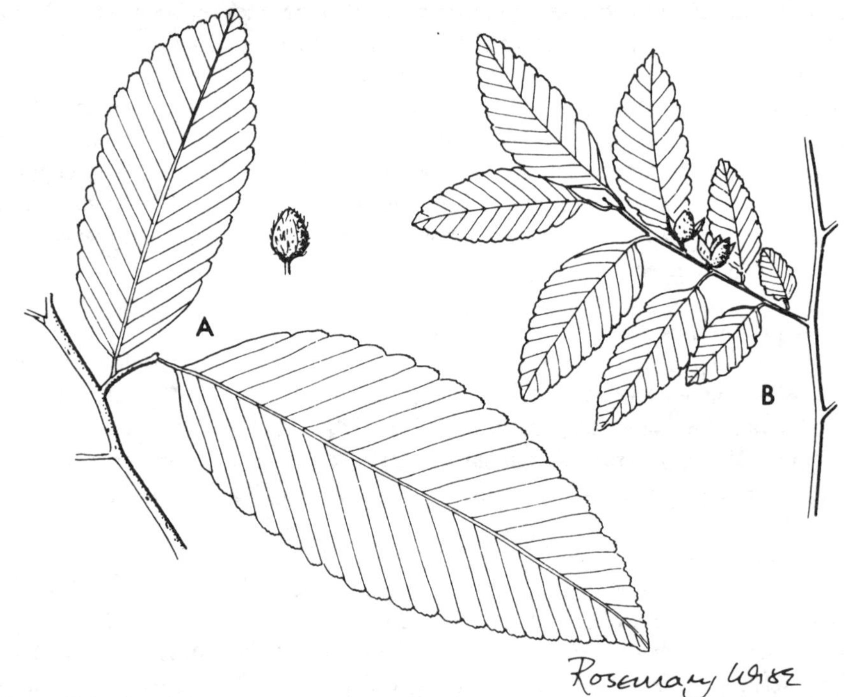

Figure 16. Southern, or Chilean beeches, A) *Nothofagus procera*, B) *Nothofagus obliqua*.

PICEA ABIES (L.) Karsten **Norway spruce**

There are 34 species of *Picea* which are confined to the cooler parts of the northern hemisphere.

Norway spruce has a useful place on wet sites, at middle and lower elevations, in regions which have too little rain for Sitka spruce. It is also one of the better conifers for planting in mixture with broadleaved trees because it is not too fast growing.

Origin and introduction

The natural distribution ranges from the Pyrenees, Alps and Balkans, northwards to south Germany and Scandinavia and eastwards through the Carpathians and Poland to western Russia, where it merges with *Picea obovata*. It is a high mountain species in central Europe and a lowland tree in northern Europe. It was probably introduced to Britain before 1500, but was a native species in the last interglacial period.

Climatic requirements

Norway spruce is a very accommodating species with no particularly marked climatic limitations in Britain, except those imposed by exposure, which it will not tolerate nearly as well as Sitka spruce. It thrives where rainfall exceeds 750 mm. It can therefore be used in drier regions than Sitka spruce but not at such high elevations. Though more hardy to spring frosts, it can nevertheless suffer very badly. It will not tolerate salty winds or industrial pollution.

Site requirements

The tree does best on moist, even moderately waterlogged rushy land of medium to high fertility, including heavy clays and the less acid peats. If the site is too dry, it tends to suffer from crown dieback (Day, 1951). Drought-prone sites should be avoided, especially in eastern Britain, and on calcareous soils the species suffers from chlorosis caused by iron deficiency. On sites dominated by heather, Norway spruce is unable to compete for nitrogen.

Other silvicultural characteristics

Norway spruce always grows slowly for the first year or two after planting. It will thrive on sites where deep rooting is not possible but the penalty for this is that it becomes susceptible to windthrow. The tree is moderately shade-bearing when young and can be used for underplanting, if freed early enough. It recovers well from underthinning. Because seed production is poor, natural regeneration is comparatively rare.

Norway spruce is a very valuable and commonly used tree for use in mixtures with hardwoods, especially oak, with which it has a compatible growth rate. Its value in this respect is often all the more important because some of the spruces can be sold as Christmas trees in many parts of the country, providing an attractive and very early financial return.

Provenance

In general, southern European provenances, from Poland, Romania, Czechoslovakia and Bulgaria, grow best in Britain. These combine fast growth with late flushing times (Lines, 1987). Many seed-bearing British stands are of French, Alpine and Scandinavian origin. They grow slowly and tend to flush too early and so are sensitive to damage by spring frosts. Seed should not, therefore, be collected from them. The only exception is seed for Christmas trees, where some south-west German sources are very suitable (Pearce, 1979).

Flowering, seed production and nursery conditions

The tree flowers in May, seeds ripen and are usually ready for collection in October, and are dispersed naturally between October and April. Viable seed is rarely produced in any quantities in Britain, largely because it does not mature properly. The earliest age at which the tree bears seeds is 30 to 35, but the best crops are usually between 50 and 60 years. There are about 195,400 seeds/kg (range 103,600 - 218,300), of which 110,000 are normally viable. For nursery purposes seed is usually sown in mid to late March. It requires no pre-treatment unless it is dormant, in which case it should be stratified for three weeks.

Area and yield

127,000 ha or 6 per cent of the forest area of Britain has crops of Norway spruce, making it the sixth most common species. Mean yield classes range from 10 to 13 m^3/ha/year in different regions (Nicholls, 1981), with a maximum of 22. For equivalent yield classes, rotations of Sitka spruce are around 15 years shorter, and on sites where both will grow, Sitka spruce is usually more productive. If a choice between the two species exists, it is therefore rare to pick Norway spruce.

Timber

Details are given with the description for Sitka spruce. The average density of the wood of Norway spruce, at 15 per cent moisture content, is about 470 kg/m^3.

PICEA SITCHENSIS (Bongard) Carrière **Sitka spruce**

Sitka spruce is probably the best adapted tree for growing in the wet, upland parts of the UK. It is highly productive, tolerant to severe levels of exposure, and will grow well on a wide variety of sites. It has become by far the most widely planted, and hence important tree in British forestry, occupying some 24 per cent of all forest land.

Origin and introduction

The natural range is rather narrow (c. 80 km wide) but very extended in the "fog belt" of the Pacific coast of North America, from Alaska to California, spanning over 22° of latitude. It is largely restricted to west-facing slopes, seldom above 900 m, but it extends up river valleys. Its inland distribution depends upon the penetration of fog banks. The largest Sitka spruce occur in the Queen Charlotte Islands and the Olympic peninsula of Washington, where the trees grow 80 m tall. It was introduced to Britain in 1831.

Climatic requirements

In its native range, the climate is maritime, with abundant atmospheric moisture. In Britain Sitka spruce rarely does well where annual rainfall is below 900 to 1,000 mm. It is therefore well suited to the higher elevations in the west and north. The dry east and south should be avoided. One of Sitka spruce's outstanding attributes is its ability to grow well, without becoming deformed, in regions of high exposure: it is better in this respect than any other common

Figure 17. A) Norway spruce, *Picea abies* B) Sitka spruce, *Picea sitchensis*.

conifer. The upper limits of planting are about 600 m on sheltered sites and 150 to 300 m lower in exposed western areas. The tree is not tolerant of pollution.

Most provenances of Sitka spruce are susceptible to damage from spring frosts when young, particularly the southerly ones from Washington and Oregon. The latter are more prone to damage from early autumn frosts as well. Cannell and Smith (1984) have shown that young trees planted in the British uplands are likely to be subjected to a damaging spring frost once every three to five years. In most places they are sufficiently fast-growing to get above the frost level quickly enough for this not to be a serious constraint on planting. However, damage in frost hollows is likely to be more severe and much more frequent, and Sitka spruce will not survive if planted in them. The species is also susceptible to frosts which occasionally occur in the autumn, before they have hardened off sufficiently for the winter (Cannell *et al.*, 1985). Such damage is much less frequent than spring frost damage, occurring on average every eight to eleven years.

Site requirements

Sitka spruce is very accommodating. It grows well on drained peats, is highly productive on gleys and does best on deep freely drained soils in suitably high rainfall areas. The tree is quite site-demanding and unless a planting area is at least moderately fertile, nutrients have to be added, particularly phosphate. It is regarded as a "high input" species in comparison with, for example, lodgepole pine. Dry and very shallow soils should be avoided: if these are in the uplands lodgepole pine or *Abies procera* should be planted. Like Norway spruce, and several other conifers, the tree will not grow in competition with heather because it is unable to compete for nitrogen (Handley, 1963): if heather-covered sites are to be planted with Sitka spruce, it is necessary either to kill the heather with a herbicide, or to provide enough fertiliser nitrogen to enable it to close canopy.

Other silvicultural characteristics

Sitka spruce can thrive on wet soils where deep rooting is not possible, though windthrow is a constant and serious risk, and a frequent occurrence, on exposed sites. It is light-demanding and cannot be used for underplanting. Natural pruning is rather slow which makes young plantations particularly impenetrable and prickly. It is unfortunate that the most widely planted tree in Britain should also be one of the most unpleasant in this respect. Very deep accumulations of fallen needles are to be seen on most upland acid sites where decay and consequent nutrient cycling is poor. This often leads to nutrient deficiencies but Sitka spruce responds well to fertilizers, even after many years of check.

As first rotations come to an end, as they have been doing increasingly since the 1970s, it has become apparent that natural regeneration is reliable on many sites on the borders of England and Scotland, and in Wales and elsewhere. It is best in places where crop stability is good, and rotations long enough for the trees to produce large amounts of seed. In many of these areas re-establishment is almost entirely by natural regeneration, and this seems likely to become a common method of replacement in the future, except on sites where new provenances, or genetically improved strains of Sitka spruce are intended.

The species is susceptible to attack by honey fungus, and therefore not usually recommended after a broadleaved crop, and is more susceptible than many conifers to *Heterobasidion annosum.* It is defoliated periodically and quite seriously by the green spruce aphid, *Elatobium abietinum* (Carter, 1977; Carter and Nicholls, 1988). It is much more seriously affected by this aphid than Norway spruce. There are also, as yet, unknown threats from the great spruce bark beetle, *Dendroctonus micans*, which was first discovered in Britain in 1982. Sitka spruce is less palatable to deer than most conifers, but it can nevertheless be severely damaged if there is no other choice.

Flowering, seed production and nursery conditions

The tree flowers in May, most seeds are ripe by September, when they can be collected, and are dispersed naturally between October and the spring. The earliest age at which the tree bears seed is 30 to 40 years, but reliable seed crops are usually at three to five yearly intervals after the age of about 50 years. There are about 463,000 seeds/kg (range 341,700 - 881,800), of which 320,000 are normally viable. Sowing times and seed treatments for nurseries are the same as for Norway spruce.

In the 1980s there was a small, but increasing use of improved Sitka spruce, produced by micropropagation techniques.

Provenance

There is a cline of decreasing vigour and increasing resistance to spring frost damage from Oregon to Alaskan seed sources, possibly with some ecotypic variation in the Queen Charlotte Islands (Lines, 1979b) where vigour and frost resistance are better. The choice of provenance must attempt to compromise on these two characteristics. In general seed from the Queen Charlotte Islands satisfactorily combines hardiness with adequate growth rates and often gives the best growth on all sites. More southerly provenances can be used in milder parts of Britain, and more northerly ones from Alaska may have a place on very exposed or elevated sites (Lines, 1987).

Area and yield

Sitka spruce is planted on 526,000 ha (24 per cent) of the forest land in Britain, making it by far the most common species. Rates of planting in recent years have approached 80 per cent. Mean yield classes range from 10 to 14 m^3/ha/year in different regions (Nicholls, 1981), and the normal maximum is about 24.

Timber

The wood of spruces is light in weight, non-resinous and rather coarse textured. On drying it is liable to twist. The timber of both Norway and Sitka spruces is preferred for pulping processes. It also provides material for light-weight types of particle board. The potential market for sawn timber is large, but rapid early growth, a tendency to spiral grain and, in the case of Sitka spruce, the difficulty of avoiding a rough finish, means that only a small proportion is suitable for joinery and other high class structural work. Grading is therefore very important. The wood of Sitka spruce is close to the lower limit of strength required for many structural purposes. Any treatment which causes the core of juvenile wood to increase in size is in danger of reducing the strength below an acceptable level. Hence much attention is paid to spacing with this species. There is also scope for selecting and breeding trees with high density wood (Savill and Sandels, 1983). The average density of the wood at 15 per cent moisture content ranges between 380 to 450 kg/m^3. Suitably graded material is satisfactory for many purposes: trussed rafters, internal framing, partitioning, etc, but it should be protected against rot and insect attack. This is difficult as the heartwood is naturally resistant to pressure treatment with preservatives.The wood of slowly-grown Norway spruce is resonant, and because of this it is used for making parts of violins and cellos.

PINUS L. — Pines

There are some 93 species of pines, including both temperate, sub-tropical and tropical species. Many of the most productive timber producing species in the world are in this genus, and they include some of the most widely planted species. *Pinus sylvestris*, the Scots pine is the only species native to Great Britain, and is one of only three native conifers.

PINUS CONTORTA Dougl. — Lodgepole pine

Origin and introduction

The range of *Pinus contorta* extends from south-east Alaska and interior Yukon in the north, to Baja California (Mexico) in the south and extends eastwards as far as the Black Hills of South Dakota (33° of latitude and 33° of longitude). The species is most abundant in the northern Rocky Mountains and Pacific coast region. It is found from sea level to about 3,600 m and covers a huge variety of climatic and soil conditions, which has led to some sub-division of the species. Lodgepole pine was first successfully introduced to Britain in 1853.

Climatic requirements

Pinus contorta grows well under very diverse, but generally upland conditions. Appropriate provenances are resistant to winter cold, spring frosts, salt-laden winds and air pollution, and will withstand great, even extreme exposure. The species is often planted in the uplands as a last resort where no other tree will thrive because of severe climatic and site conditions.

Site requirements

Some provenances will grow fast on the poorest soils such as deep acid peats, hard boulder tills and upland heaths. Performance can be outstanding compared with other species on such sites, provided there is adequate phosphorus in the soil. The tree is much more tolerant of competition from heather than Sitka spruces, and if this is likely to be a problem, lodgepole pine is often a natural choice, at least for a first rotation. It is a coarse and unattractive tree on good lowland sites, where several other species are much more productive.

Other silvicultural characteristics

Lodgepole pine is commonly planted at high elevations on the poorest western and northern soils because it is relatively undemanding, and will grow well with only low inputs of fertilisers. It has the reputation for drying out peat in some areas

(Pyatt and Craven, 1979). Natural regeneration can be good, especially on burnt sites.

Lodgepole pine used to be widely planted as a nurse, to provide shelter and to suppress weeds, usually in mixtures with Sitka spruce. However, since the late 1960s, this practice has been largely abandoned as establishment techniques for spruce have improved. Difficulties were also quite frequent in that the pine grew so much faster than the spruce that the latter was eventually suppressed and killed. Some of these mixtures have led to the discovery of the most remarkable additional nursing effect on Sitka spruce on some peaty sites deficient in available nitrogen. This was first described by O'Carroll (1978) and has been much more widely observed since, especially with Alaskan origins of lodgepole pine (Lines, 1987). Spruces in such mixtures often grow at rates which are many yield classes higher than in unmixed crops. The mechanisms involved are not yet understood though they clearly involve improving the nitrogen nutrition of the spruces.

In spite of its virtues, lodgepole pine has many serious problems too, some caused by stress brought on by the extreme environments in which it tends to be grown. It is vulnerable to damage by many organisms: it is attacked by the pine beauty moth, *Panolis flammea* (Stoakley, 1979), pine sawfly, pine looper moth (Bevan and Brown, 1978), pine shoot beetle (Bevan, 1962), badly by deer, and on infected sites by *Heterobasidion annosum*. It is susceptible to snowbreak and windthrow on wet soils, and then (unlike spruces) the timber decays within about a year, so it must be salvaged quickly which can be difficult if windthrow is extensive. The most vigorous provenances are very coarse and suffer from basal sweep. These problems lead some foresters to prefer leaving areas unplanted if the alternative is pure lodgepole pine (Davies, 1980).

Provenance

Three main interfertile races are usually recognised as subspecies. They differ markedly in ecology and morphology and can lead to enormous differences in growth and yields of timber. Subspecies *contorta* grows along the Pacific coast from British Columbia to California in bogs, on sand dunes and on the margins of pools and lakes. It is short, shrubby and of poor form but is much the fastest growing in Britain. Subspecies *latifolia* is found in the intermountain systems from central Yukon to east Oregon and south Colorado, and subspecies *murrayana* grows mainly in the Cascade and Sierra Nevada mountains of Oregon and California (Critchfield, 1957). The latter two subspecies are well formed, slender, tall, straight trees, but are much less productive.

A constant dilemma has been to decide how far to compromise on form while retaining vigour. In the Republic of Ireland the most vigorous provenances of subspecies *contorta* from Washington and Oregon are almost exclusively used (O'Driscoll, 1980), while in Britain slower growing and more northerly coastal provenances are favoured, besides some inland provenances of *latifolia* and intermediate ones between the two. The north coastal provenances possibly have an advantage in that they are less susceptible to damage by the pine beauty moth

(Leather, 1987). The correct choice of provenance is complex, but a detailed guide is provided by Lines (1985a). In the future, interprovenance hybrids may offer possibilities for combining vigour and good form.

Flowering, seed production and nursery conditions

P. contorta ssp *contorta* flowers in May and June and seeds ripen 18 months later, between December and March. They are dispersed in the spring. The earliest age at which the tree bears seeds is at 10 to 20 years but the best seed crops are usually at intervals of two or three years after the age of 30 to 40. Seed is ready usually collected in October and November. There are about 297,600 seeds/kg (range 244,700 - 363,800), of which 270,000 are normally viable. In nurseries, the seed should be sown between late February and mid March, if not stratified or pre-chilled, otherwise in late March (Aldhous, 1972).

Area and yield

Lodgepole pine comes second to Sitka spruce in the extent of recent planting which is a reflection of the poor quality of the land. About 127,000 ha of the total forest area (6 per cent) is planted with the species. Mean yield classes range from 6 to 10 m^3/ha/year (Nicholls, 1981), with a maximum of about 14 for coastal Oregon and Washington provenances, though yield classes as high as 18 are commonly found in Ireland.

Timber

Except that it contains a greater proportion of heartwood, lodgepole pine is otherwise a very similar and satisfactory alternative to Scots pine, though it is more resistant to penetration by preservatives. It has a straight grain, low distortion on seasoning, high stability and a comparatively uniform texture. A smooth finish can be obtained, even on wide-ringed timber and it should make a suitable joinery wood. Its value may, of course, be reduced considerably by the poor form of many trees of ssp *contorta*, and by knots. The average density of the wood at 15 per cent moisture content is about 470 kg/m^3.

PINUS MURICATA D.Don — Bishop pine

Origin and introduction

Bishop pine is found in more than 15 scattered colonies along the coast of California and on the nearby Cedros Islands. It was introduced to Britain in 1846.

Climatic requirements

These are uncertain but the tree is very close to, or slightly beyond the limit at which it can safely be grown in Britain. Its requirements appear similar to those of *Pinus radiata*, though *P. muricata* is slightly less demanding. It is largely confined to low elevations in south-west England but may be reasonably safe south of a line from the Dee to the Wash according to Everard and Fourt (1974). Quite large areas have been satisfactorily established by the Forestry Commission at Wareham forest in Dorset.

Site requirements

These are not really known. Bishop pine is believed to be more tolerant of poor soils than *P. radiata*, but calcareous soils should be avoided.

Other silvicultural characteristics and yield

The tree is often multi-stemmed and very coarse. It needs singling and pruning early, often by age three or four, and later high pruning. The species is one of the closed-cone pines, and cones are retained on the trees. It often suffers badly from the pine budmoth, *Racinonis buoliana.*

Yield classes of 20 m^3/ha/year can readily be achieved, which is the main source of interest in this otherwise unattractive species.

Provenance

Within its natural range there is a southern "green" form and a northern "blue" one. The latter is better shaped and appears to be more frost hardy (Tuley, 1979).

PINUS NIGRA J.F. Arnold **Black pine**

Pinus nigra has a discontinuous distribution in central Europe and the northern Mediterranean region from southern Spain to Turkey and the Crimea (35° to 49°N, and 3°W to 34°E). It is predominantly a mountain tree but also occurs at sea level along the shores of the Adriatic. Four varieties are now generally recognised:

var. *caramanica*	Crimean pine
var. *cebennensis*	Pyrenean pine
var. *maritima*	Corsican pine
var. *nigra*	Austrian pine

Only var. *maritima* is used as a timber tree in Britain. It originates from Corsica, southern Italy and Sicily and was introduced in 1759; var. *nigra*, from Austria, central Italy and the Balkans will also grow.

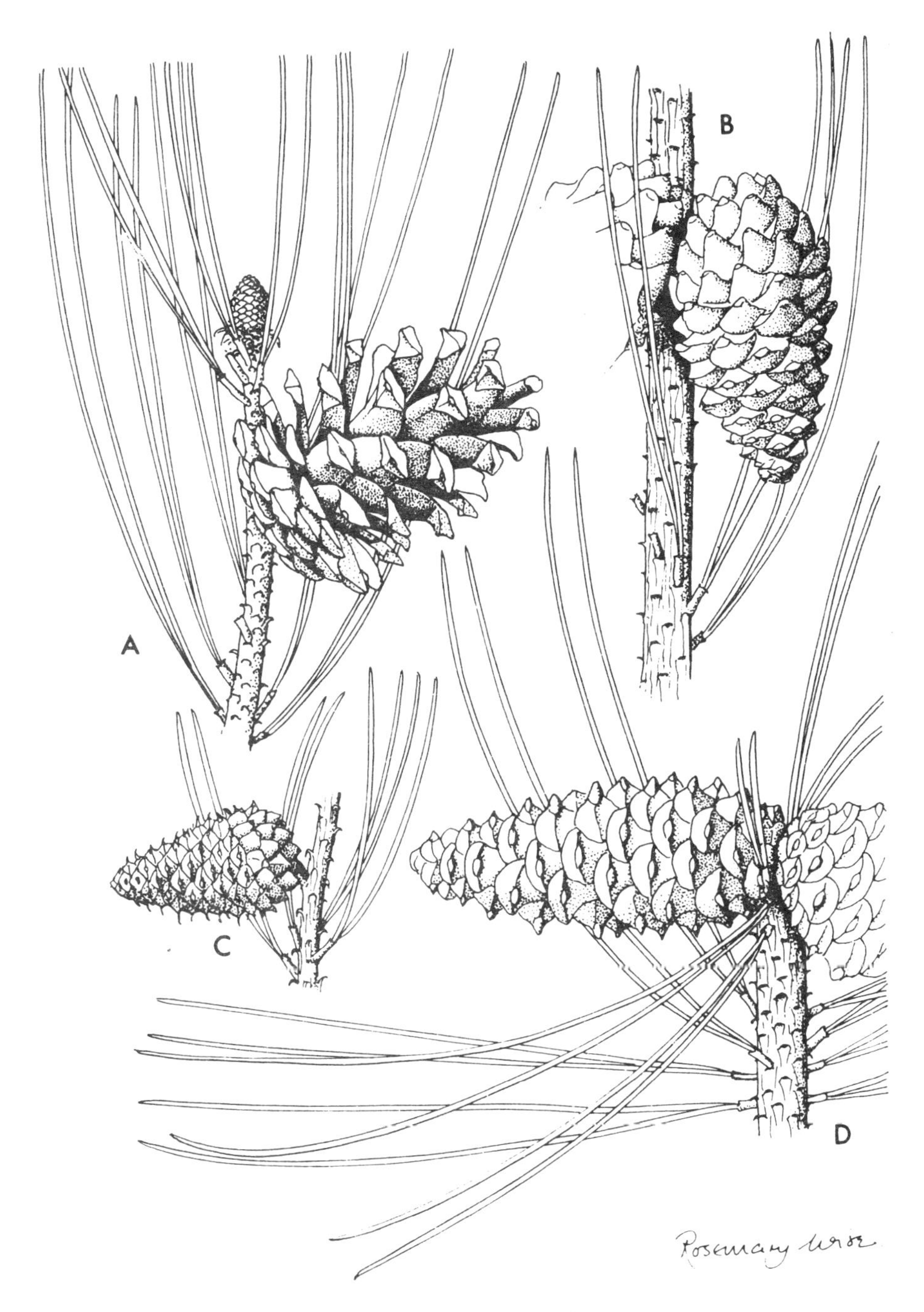

Figure 18. A) Austrian pine, *Pinus nigra* var. *nigra*;
B) Bishop pine, *Pinus muricata* C) Lodgepole pine, *Pinus contorta*:
D) Corsican pine, *Pinus nigra* var. *maritima*.

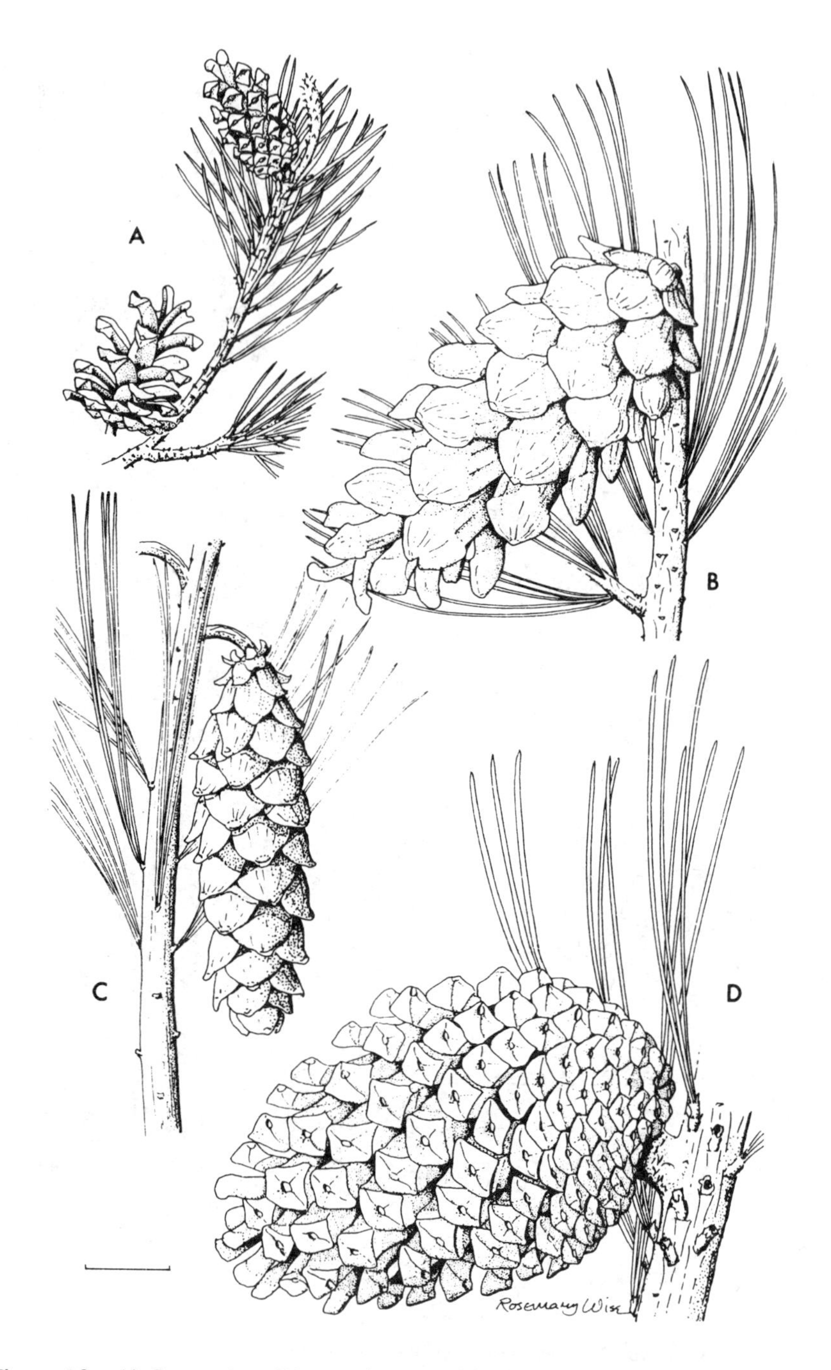

Figure 19. A) Scots pine, *Pinus sylvestris*; B) Macedonian pine, *Pinus peuce*; C) Weymouth pine, *Pinus strobus*; D) Monterey pine, *Pinus radiata*.

PINUS NIGRA var. MARITIMA (Ait.) Melville
Corsican pine

Corsican pine is regarded as the most profitable and suitable conifer for planting in much of lowland Britain, particularly on light soils in the east of the country.

Climatic requirements

It thrives in the south and the Midlands of England, the coastal fringe of south Wales and along the east coast of Scotland. These are all districts of low summer rainfall, with high summer temperatures and duration of sunshine. In wetter, cloudier districts of the north and west it is usually fatally attacked by a die-back fungus *Sceleroderrus lagerbergii* (Read, 1967). For this reason uplands should normally be avoided, but if planted, it should be confined to warmer south and west-facing slopes, brashed early and thinned heavily in an attempt to prevent infection by improving air circulation. As a rough guide, Lines (1987) suggests elevations of no more than 300 m in south-west England as being safe. Wetter upland sites, over 150 m in north England and lower further north, carry some risk of dieback. In milder parts of the lowlands it grows well on sites which also suit Scots pine. It does best in areas of low rainfall and tolerates smoke better than most conifers. It will also grow better than Scots pine on coastal sand dunes subjected to salty winds. Corsican pine can be very difficult to establish in places where frequent and severe late spring frosts occur. The risks of damage can be minimised by planting on ploughed ground or otherwise maintaining bare soil round the plants, as a means of preventing temperatures from falling too low.

Site requirements

Corsican pine grows well on a wide range of mineral soils, but with least risk on sands and heavy clays in the midlands, south and east of England. Though more successful than Scots pine on chalky soils, it does not thrive on them.

Other silvicultural characteristics

Conventional bare-rooted plants are notoriously difficult to establish. They are sensitive to bad handling and, especially to the roots drying out. Very high, and even complete losses are common, and this has encouraged the increasing use of container-grown (Japanese paper-pot) plants. More recently, trials with undercutting nursery seedbed stock has proved encouraging. To ensure reasonable survival with transplants or plants in pots, planting should be carried out as late as possible in the spring. Very little growth occurs in the first year or two after planting, so that careful weed control is necessary to prevent smothering.

The tree is strongly light-demanding. Though the form of trees is generally very good, branching can be heavy and the species does not self-prune well. Timber quality can therefore be considerably improved by pruning, though this is seldom done. Trees are not easily damaged by grass fires when in the pole stage or later. One important attribute is that Corsican pine is seldom attacked by rabbits and hares, unlike Scots pine which is often planted in the same areas. Like most pines, the species is more susceptible to attack by *Heterobasidion annosum* than many conifers. At Thetford Forest, stumps of clear-felled crops are removed from the ground in the worst infected areas in an attempt to reduce the levels of infection to the succeeding crop. Corsican pine is rarely attacked by the pine tortrix (*Rhyacionia buoliana*), or the resin top disease, both of which can cause severe damage to Scots pine in East Anglia. It is sometimes susceptible to the aphid *Schizolachnus pineti* and the pine looper moth, *Bupalis piniaria*.

Flowering, seed production and nursery conditions

The tree flowers in late May and early June; seeds ripen between December and March 18 months later, and are dispersed in the spring. They are normally collected in January. The earliest age at which the tree bears reasonable crops of seed is 20 to 30 but the best seed crops are usually at three to five yearly intervals between the ages of 60 and 90, which tends to be after the normal commercial rotation age. This is one reason why natural regeneration is comparatively rare. There are about 57,300 seeds/kg (range 30,900 - 86,000), of which 55,000 are normally viable. Most seed is produced in the south-east and east of England. If required for the nursery, it should be sown between late March and early April. No pre-treatment is required (Aldhous, 1972).

Provenance

Seed from Corsica grows best in Britain (Lines, 1987) and trees from this source have excellent form. Many British plantations of Corsican origin exist from which seed can be collected, particularly in East Anglia (Kennedy, 1974).

Area and yield

Crops of Corsican pine are planted on 47,000 ha (2 per cent) of forest land. Mean yield classes range from 9 to 13 m^3/ha/year (Nicholls, 1981), with a maximum of about 20. The potential mean should be 18 or more on a wide range of sites (Fourt *et al.*, 1971), but phosphorus is often limiting. In suitable areas such as the Brecklands of East Anglia it is always much more productive than Scots pine, which it is replacing.

Timber

Corsican pine makes heartwood much more slowly than Scots pine, but the timber is otherwise very similar and can be used for the same purposes. The wood is stable, with intermediate strength: stronger than spruces but weaker than larches. It is suitable for use in building, roofing, flooring and interior framing provided the correct grade is selected. Large knot whorls are potential zones of weakness. The timber is not durable, but is easily treated with preservatives and is therefore useful where a high standard of preservative treatment is needed: sleepers, poles and fences will last 50 years or more if well treated. Freshly felled logs are very susceptible to attack by blue stain fungi and should therefore be extracted, converted and dried with a minimum of delay. The average density of the wood at 15 per cent moisture content is about 510 kg/m^3.

PINUS NIGRA var. NIGRA — Austrian pine

Austrian pine is much more at home on exposed chalk and limestone than Corsican pine. It is one of few conifers with this attribute and it grows well even where the soil is dry and shallow. It can be useful as a nurse for beech on such sites. It will not thrive on wet, heavy soils, but tolerates sea winds and industrial pollution very well. It has a coarse and poor form, and largely because of this it is virtually useless as a timber tree.

Austrian pine is only worth planting as a short-term nurse (for example for beech on calcareous soils), and where limestone, salty winds or pollution rule out other conifers, and then only as a shelterbelt or for amenity.

PINUS PEUCE Grisebach — Macedonian pine

Forestry in the uplands of Britain is often regarded as being too dependent upon too few species. There has been concern for some time to find possible alternatives, particularly to Sitka spruce and lodgepole pine. *Pinus peuce* is one which is being considered.

Origin and introduction

The species is a five-needled pine, native to the Balkans: southern Yugoslavia, western Bulgaria and Albania, where it occupies a total area of no more than 30,000 ha, ranging from 1,100 to 2,300 m in elevation. It was introduced to Britain in 1864, having only been discovered in 1839.

Climatic and site requirements

Very little experience exists with the tree in Britain, but such as it is, together with a full account of the tree's performance in its natural habitat, has been given by Lines (1985a). It appears likely to grow well on a wide range of soils, including peats in such inhospitable places as the central Highlands of Scotland. It withstands exposure well and also atmospheric pollution.

Other silvicultural characteristics

Difficulties with *P. peuce* arise in the nursery and establishment phases, which is probably why it has received so little attention up to now. Early growth is very slow. Until the fifth or sixth year the trees have a dense, bushy form before making strong vertical growth. The bark remains thin for up to 30 years, and so is potentially susceptible to being stripped by deer.

Very slow early growth is also found in the American *P. palustris* (longleaf pine). It goes through a period of slow growth called the "grass stage" which is believed to be under strong genetic control. Although the length of time individual seedlings remain in this stage is influenced by the environment, it can last for 25 years, and often for 15 years. Reasonable stands will reach breast height at about eight years. Generally seedlings remain in the grass stage until they reach 2.5 cm at the root collar and they invariably begin height growth upon reaching that size. The control of competition is a major factor in stimulating fast diameter growth: if it is very good, height growth can begin at the end of the second year (Walker and Wiant, 1973). It is possible that a similar mechanism may operate in *P. peuce*.

Unlike *P. strobus*, another five-needled pine, it is resistant to attacks from the blister rust *Cronartium ribicola*, and to several other likely pests and diseases of pines. In fact it seems likely to be a remarkably hardy and healthy tree in Britain.

Seed production and nursery conditions

Seeds are fully ripe in September, but are often completely stripped, at least in small plots, by squirrels in August. Lines (1985b) states that one of the major disadvantages of Macedonian pine is the tendency to poor or delayed germination, which is partly due to incomplete embryo development. Usually some seeds germinate in the first spring but many do not germinate until the second.

Yield

There is increasing evidence that the species will be a high volume producer in Britain compared with other pines. A feature is that in comparison with other species, basal area growth can be up to 50 per cent greater for a given height. This makes it of interest where the risk of windthrow is high.

Timber

Preliminary studies in Britain indicate that an important attribute of the wood is its stability compared with other common coniferous timbers, though its strength is poor. Its density at about 12 per cent moisture content is about 350 kg/m^3, which is considerably lower than Scots pine.

PINUS RADIATA D.Don

Monterey pine; radiata pine

This species is of very minor importance in Britain but interest in it arises because of its very rapid rate of growth in the few places where it survives.

Origin and introduction

Pinus radiata is native to three small areas in the "fog belt" on the coast of California, round Monterey and Cambria, between 35° and 37°N, and also on three small islands off the coast; Santa Cruz, Santa Rosa and Guadaloupe. It has no economic importance in its native range, but as an exotic it is probably the most widely planted of all trees, and is the basis of large timber industries in New Zealand, parts of Australia, and Chile. It was introduced to Britain in 1833.

Climatic requirements

P. radiata is probably beyond the limit of where it can safely be grown, even in the most favoured parts of southern England. Its climatic requirements are not well understood but it may be more hardy to low winter temperatures than generally believed: tolerance to frosts of -6°C in summer and -14°C in winter have been found in New Zealand (Menzies and Chavasse, 1982). It probably needs a long, warm growing season. Green (1957) considered it required an "accumulated" temperature above about 1,400 day°C (i.e. the sum of the mean monthly temperatures above 6°C x days in the month). Much of southern England has these temperatures, but recent experience suggests that even higher levels may be needed, possibly over 1,700 day°C, such as are found in parts of Dorset, Somerset, Devon, Hampshire and Cornwall at low elevations. Rainfall is not limiting. In unsuitable climates trees of any age may suddenly go yellow and needle retention is reduced; they may die quite suddenly.

Site requirements

P. radiata has grown well on deep, dry and infertile sandy soils in the south of England, and grows on loams and clay loams. Very wet, and shallow calcareous soils over chalk should be avoided.

Other silvicultural characteristics and yield

Growth is rapid where the tree survives and this is the cause of the interest in the species. Yield classes of 18 to 22 m^3/ha/year are common in south-west England with maximum mean annual increment occurring before age 30. The average density of the wood at 15 per cent moisture content is about 480 kg/m^3. The tree is coarsely branched, so pruning is necessary, and is attacked by the pine shoot moth, *Rhyacionia buoliana*.

PINUS STROBUS L. — Weymouth pine

Weymouth pine, like *Pinus peuce*, is a five-needled pine. It is a species with some potential in Britain, but it is not planted because of fatal infections, common in five-needled pines, from the blister rust *Cronartium ribicola*, which was first noticed in 1892. The main alternate host of the rust in the UK is the blackcurrant, *Ribes nigrum*, and the risk of infection has precluded the use of the pine. Some work is being done in Germany and elsewhere on breeding rust-resistant strains.

The species is native to, and widespread in, north-eastern North America and is found occasionally in Mexico and Guatemala. It grows at low elevations in the northern part of its range and in the mountains in the south, in regions of cool, humid climates. It was introduced about 1705 and named after Lord Weymouth.

Its site requirements are said to be quite exacting (Elwes and Henry, 1906): good loamy hardwood sites are needed or deep sandy or sandy loam soils.

The timber tends to be less dense than that of many pines, at about 420 kg/m^3.

PINUS SYLVESTRIS L. — Scots pine

Scots pine is one of three native conifers, and the only one of any commercial significance in Britain. Its natural range extends from Spain, across the whole of northern Europe and Siberia, almost to the Pacific ocean. Within Great Britain, *P. sylvestris* var. *scotica* is the native tree of the Highlands of Scotland (Clapham *et al.*, 1985).

Climatic requirements

Scots pine is an adaptable species but does best in the drier eastern districts. It does not grow well where exposure is excessive and is not therefore a tree for high elevations unless there is adequate topographic shelter.

Site requirements

Scots pine grows best on light, non-calcareous soils, especially sands, gravels and other well drained sites at lower elevations. Peaty ground is best avoided. It has a relatively short life on calcareous soils.

Other silvicultural characteristics

Lines (1987) described Scots pine as a typical light demanding, pioneer species. It is very frost-hardy and grows rapidly when young. It will, however, regenerate in open stands, but if the regeneration is to succeed, it must be released within ten years. If the canopy opening is delayed, the regeneration will not respond. It is often planted in mixture with broadleaves and is useful as a nurse for this purpose because it does not grow too fast, especially with beech on frost-prone chalk and limestone soils. Where it is truly at home, such as in parts of the New Forest, natural regeneration can be prolific and invasive. The tree does not self-prune well and to obtain clear timber without loose, dead knots, pruning is necessary.

It suffers from resin top disease (Pawsey, 1964), particularly in north-east Scotland, and from the rust, *Peridermium pini.*

Flowering, seed production and nursery conditions

The tree flowers in May and June; seeds ripen in September and October of the following year, and are dispersed between December and March. The best seed crops are usually produced at three to five year intervals after the age of 60. Seed is normally ready for collection in January. There are about 165,300 seeds/kg (range 74,500 - 244,700), of which 140,000 are normally viable. The treatment of seed in nurseries is the same as for Corsican pine.

Provenance

Seed for use in Scotland and north-east England comes from Forestry Commission seed orchards from which substantial amounts of improved seed are now available.

Area and yield

In terms of the total area planted, Scots pine comes second to Sitka spruce, with 241,000 ha (11 per cent of forest land,) but its present area gives a false impression of current levels of planting, which are almost negligible. Its rather slow growth and long rotations compare badly with more exacting species. It is, for example, being rapidly replaced by Corsican pine in eastern England. Mean yield classes range from 8 to 11 m^3/ha/year in different parts of Britain (Nicholls, 1981), with a maximum of 14.

Timber

The wood of Scots pine (known in the trade as deal or redwood), has been the standard utility timber of northern Europe for generations. The average density of the wood at 15 per cent moisture content is about 510 kg/m^3. It combines adequate strength with light weight and is easy to nail and work. Clear narrow-ringed wood can be of excellent quality for joinery and most is suitable for building. It is easily treated with preservatives and very useful where a high standard of treatment is required, as in railway sleepers and fencing. Clear material can produce an excellent veneer.

POPULUS L. Poplar

Poplar cultivation is very lucrative in parts of Europe and other places well suited to their growth, but in Britain poplars are planted mostly for screening, shelter or ornament, rather than for the production of timber. Poplars, their cultivation and uses, have been described in detail by Jobling (1990).

Origin

About 35 species occur in the northern hemisphere, in the boreal and temperate zones between the subarctic and subtropical regions. The aspen, *Populus tremula*, and black poplar, *P. nigra* var. *betulifolia*, are certainly native to Britain, while the grey poplar *P.* x *canescens* is thought to be native to southern England. There are numerous cultivated hybrids, varieties and clones which are propagated vegetatively, from cuttings or sets. Those in current use are various crosses between known individuals of the European *P. nigra* and the two north American species *P. trichocarpa* and *P. deltoides*.

Many poplar species interbreed freely, and the presence of numerous natural hybrids presents the taxonomist with difficulties in defining the limits of species. This situation is complicated by the occurrence of leaf dimorphism within species and evidence that some "species" are themselves intersectional hybrids. The identification of most cultivated poplars is therefore extremely difficult, and frequently impossible from morphological characteristics alone. However, Greenaway *et al.* (1991) have described the copious sticky exudate from the buds of most poplars, which consists primarily of a complex mixture of phenolic compounds. The composition of this mixture is characteristic of species, and even morphologically similar clones can be distinguished by analysis of their bud exudate using gas chromatography - mass spectrometry.

Clones

In 1985 a number of new poplar clones were introduced to Britain, having been bred at the Poplar Research Centre at Geraardsbergen, Belgium (Jobling, 1990). They are the result of nearly 20 years of disease screening, testing and selection. They differ from previous clones which have been widely planted in the UK in that they are *P. trichocarpa*/ *P. deltoides* crosses, with parents originating from similar latitudes to those of the UK. Growth rates in Belgium are significantly higher than those of the clones already in use. The Lombardy poplar is a fastigiate variety of *P. nigra* and is much planted for ornament.

Climatic requirements

Poplars are of minor commercial importance in Britain because they are near their climatic limits as forest trees. On suitable sites, they grow best in southern Europe where summers are hotter. In Britain, they must be confined to sheltered sites in lowland regions. About half of all British poplar growing is carried out in East Anglia.

Site requirements
Cultivated varieties are very exacting in their site requirements and suitable sites available for production are few. They need highly fertile, base-rich, loamy soils or rich alluvial or fen soils which are well drained and aerated, and moist even in conditions of summer drought, usually with a water table within 1 - 1.5 m of the surface. Banks of streams and the alluvial soils of river valleys are very suitable. Most poplars will grow reasonably well on a much wider range of soils, but not fast enough for commercial crops. Sites to avoid are those with acid soils, shallow soils and soils where there is stagnant, or very slowly draining water.

Other silvicultural characteristics
Poplars are strongly light-demanding and must be established at very wide spacings. They are normally planted as pure, widely spaced crops, by pit planting. Spacings of about 8 x 8 m are used in Britain for growing 45 cm diameter at breast height trees, with no thinning, on rotations as short as 22 years (Beaton, 1987). At such wide spacings, regular pruning is necessary if the wood is not to become excessively knotty. Poplars are also most intolerant of competition with weeds and grow extremely slowly when young unless almost complete weed control is practised.

There is considerable recent interest in growing poplars as short rotation biomass crops for the production of energy or industrial uses, particularly in countries which do not have large reserves of fossil fuels (Hummel *et al.*, 1988), and as a species in "agroforestry" systems (Beaton, 1987).

The most serious disease of poplar in Britain is bacterial canker caused by *Xanthomonas populi*. Infection via natural openings such as leaf or bud scale scars can lead to the death of branches and to the development of perennial cankers on the trunk. *Melampsora* rust is a leaf disease characterised by the presence of bright orange pustules on the underside of leaves. Although severely affected leaves are very conspicuous, the fact that the disease does not develop until late in the growing season means that there is little effect upon growth. The poplar leaf spot caused by *Marssonina brunnea* can also cause problems to some clones, but all the 1985 introductions have relatively high resistance to these three diseases (Potter *et al.*, 1990).

Most poplars are susceptible to bark stripping by grey squirrels. Because their roots are often very close to the ground surface, they can be troublesome near buildings, blocking drains and damaging paving.

Defoliation can be caused by sawflies and leaf beetles, and serious damage to the timber may result from the larvae of an agromyzid fly which bores long tunnels in the stem cambium. Two aphids, *Pemphigus bursarius* and *Pemphigus phenax*, migrate from black poplars (especially Lombardy poplars) in summer to the roots of lettuces, sowthistles and related species, and to carrots respectively

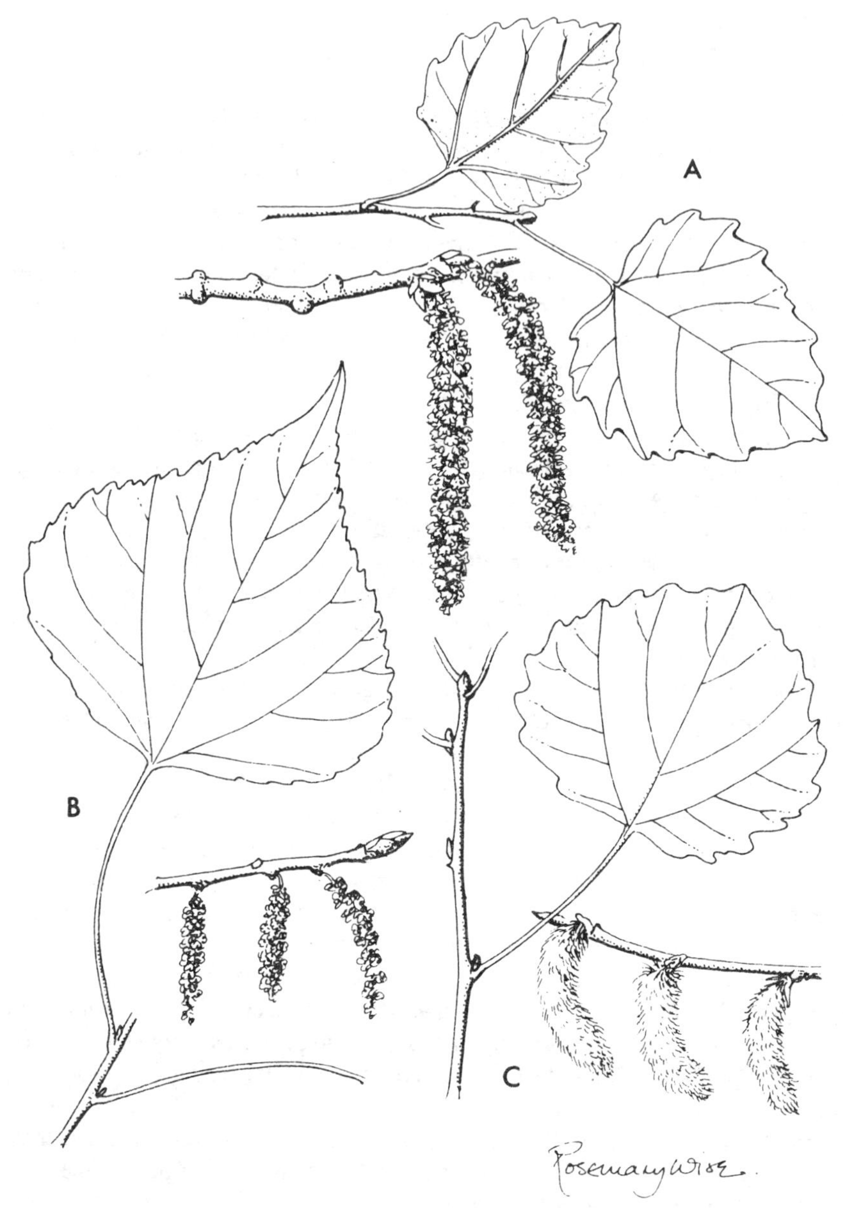

Figure 20. A) Grey poplar, *Populus* x *canescens*;
B) Black poplar, *Populus nigra* var. *betulifolia*;
C) Aspen, *Populus tremula*.

(Blackman, 1974). Quite serious economic damage can be caused if poplars are planted for shelter where such crops are grown. Another aphid, *Dysaphis apiifolia* similarly attacks celery.

Propagation

Recent breeding programmes have resistance to the fungi *Melampsora, Marssonina* and *Dothiciza* as the first step in screening, ahead of morphological and production characteristics. Most species flower in the early spring, but poplars are normally propagated from cuttings in open nursery beds. For the majority of clones these are taken from ripened hardwood cuttings, but for a few (particularly *P. tremula* and *P.* x *canescens*), root cuttings, root suckers, or leafy softwood cuttings from the current shoot growth are used. Details are described by Jobling (1990). Propagation from seed is very rare in Britain.

Area and yield

About 14,000 ha of poplars have been planted in Britain, which represent 0.6 per cent of the total forest area (Locke, 1987). Yield classes of up to 14 m^3/ha/year are normal with the older clones, but the new ones are said to be capable of 18 to 22 m^3/ha/year.

Timber

The wood of most poplars is ideal, and the best of all temperate timbers, for peeling into thick veneers. It is soft, light coloured, of low density (about 300 to 550 kg/m^3 at 12 per cent moisture content), and fine textured, but quality varies according to species, region and conditions of growth. It is valued for some special purposes because it is very tough for its weight and has the unusual attributes of bruising rather than splintering when subjected to abrasion (which makes it valuable for pallets and wagon bottoms), and of smouldering rather than igniting when violent friction is applied.

The wood is usually white and free of taints and smells and is therefore valuable for use in contact with food (e.g. vegetable crates). Green logs can be peeled cold to produce serviceable veneers from which crates, and increasingly other materials, are made. The timber is not durable and the heartwood is resistant to penetration by wood preservatives (Anon, 1980). It is, however, a good sawtimber when dry, being stable, and taking a good finish. It can be used for joinery and even some musical instruments.

Large areas of poplar were established in the 1950s and 1960s for matches and horticultural veneer, but these uses have ceased to exist in Britain because imported matches are cheaper, and plastic containers are now used for fruit. There are other uses for veneers, such as separators for loads of bricks when they are being transported, but if an industry is to develop based on poplar, sufficient supplies must be available, and this will require additional planting at the rate of something like 1000 ha/year for 25 years.

PRUNUS AVIUM (L.) L. — Cherry, or gean

The genus *Prunus* contains some 400 species, many of which are found in the northern hemisphere. It includes plums, cherries, almonds, apricots and peaches.

Two species are native to Britain. *P. avium*, the wild cherry or gean, is by far the more important, and interest in it arises from the high value of its timber and its fast early growth. Cherry is a surprisingly under-used tree. It is easy to establish and very productive for a broadleaved species, and has short rotations. It has attractive flowers and autumn colour and a much sought after timber. Why it has always been a species of such minor importance is an enigma. It might have been merely overlooked, or possibly there is still some problem which has not been recognised.

The other species is *P. padus* L., bird cherry. It is usually a very small suckering tree which grows to a maximum of 15 m tall, most commonly in northern England and Scotland on acid soils, especially in valleys and wet fen woodland. In drier areas it grows where the ground water is calcareous. The tree is very hardy, but not tolerant of exposure to strong winds. It is not considered further here.

Origin

P. avium is native to Europe, north Africa and western Asia. It occurs throughout Britain but its regional distribution is strangely patchy. It typically occurs in small groups, often on the margins of woods. This may be because of the more favourable conditions found on banks along the boundaries of woods, or possibly because it was planted there. The silviculture of cherry has been described in detail by Pryor (1985, 1988).

Climatic requirements

Cherry is essentially a lowland species, seldom being found above 300 m in elevation. It tolerates exposure badly, becoming deformed. The species is strongly light-demanding, except when very young.

Site requirements

Cherry needs a deep moist soil for good growth. It grows best on the deep soils developed in the thicker layers of drift over chalk and limestone, and on deep flushed soils on lower valley slopes. Parts of the Sussex and Wiltshire downs, the Chilterns and the Cotswolds are particularly suitable. Soils to avoid are shallow ones where the C horizon is within 40 cm of the surface, sands and wet areas.

Sites for establishing cherry orchards for fruit production are on light, well-drained soils: such soils as "brickearths" and light to medium loams where there is good under drainage. Fruiting trees do not thrive on cold, heavy clay or badly drained soils of any kind.

Other silvicultural characteristics

Cherry has strong apical dominance, and relatively weak phototropic tendencies. This usually results in the tree developing and retaining a single straight leading shoot. Stocking levels can therefore be lower than with other broadleaves, because less selection is needed of final crop trees. Planting at about 3 x 3 m is recommended. Young plants are very sensitive to competition from weeds, and full control can more than double early height growth. Cherry grows faster than almost any other species in tree shelters. The branch whorls are retained for many years after their deaths, and so will form dead knots in the wood. Hence, pruning to at least 5 m is most desirable. The tree coppices rather poorly, but suckers well.

Because cherry rotations are so similar to those of conifers, "nurses" are not needed. In fact cherry has a useful place as an early-maturing component in mixtures with other broadleaves. Height growth is very well matched to that of the larches for the first 45 or 50 years (i.e. all of a larch rotation). Cherry may suppress oak when planted in mixture, but mixtures with ash are very appropriate, the cherry being removed 10 to 15 years before the ash, or possibly at the same time.

It is rather a short-lived species, and liable to windthrow and heart rot after 60 years. There is often a race against time to achieve sawlog sizes before rot sets in. Thinnings must be heavy and regular to develop unimpeded crowns and suitable diameters early, since recovery from suppression is not good in stands over 40 years old. It is often regarded as being an ideal species for woodland edges, where crowding is not a serious problem, and the benefits of its attractive appearance are most obvious.

A virtue of cherry is that it is not damaged by grey squirrels, but deer are serious pests. It sometimes suffers from canker (*Pseudomonas mors-prunorum*). According to Roach (1985), bacterial canker is much more common in English cherry orchards than those in the drier areas of France and Italy. It is said to be associated particularly with sites where soils are deep and well drained and, possibly, where soil potassium levels are low (MAFF, 1980). An aphid, *Rhopalosiphum padi*, (which possibly occurs only on *P. cerasus*) and the latter species on apple, rowan and hawthorn migrate to the underground parts of oats, other grasses and reeds in summer (Blackman, 1974) and can cause some damage to crops. Cherry is also one host of an aphid which carries the barley yellow dwarf virus and hence it may not be wise to plant too many trees in arable areas.

Cherry itself is attacked by a number of virus diseases. It is important to try to obtain virus-free seed. Registered sources are now available in France, where the tree is also propagated by cuttings. It is also susceptible to an often fatal bacterial canker (*Pseudomonas seringae*) which is well known to growers of fruiting cherries and plums. It is characterised by the exudation of gum from the bark. The cherry black fly, *Myzus cerasi*, forms colonies in spring on the new growth and causes leaves to curl and stunting of shoots, and may result in dieback.

Flowering, seed production and nursery conditions
The tree flowers from early April to mid May; fruits ripen in June and July, and are commonly collected for seed in September. Cherry normally flowers at less than ten years old, and produces good seed crops every one to three years, especially after the age of 30. In natural conditions they are disseminated by birds. There are about 5,100 seeds/kg (range 3,200 - 6,600), of which 80 per cent normally germinates. Aldhous (1972) states that cherry seed should be kept cool, in an airtight container, between the time of extraction from the fruit and sowing or stratification. It can be sown immediately after collection, or stratified for four months and sown in March or early April.

Yield and rotations
Average yield classes of 6 to 10 m^3/ha/year are found on most sites, which is high for a broadleaved species. Rotations are only five to ten years longer than for most conifers.

Figure 21. Cherry, *Prunus avium*.

Provenance

In continental Europe, different races and even subspecies have been recognised, but no work of this kind has been done in Britain. Cherries cultivated for their fruits are derived from two species: the sweet ones are *P. avium* and the acid Morello varieties are *P. cerasus*. Duke cherries are considered to have arisen from crosses between the two (Roach, 1985).

Timber

The demand for the decorative and serviceable cherry timber greatly exceeds the supply. It is sought after for furniture in particular, and for veneer and turnery. Prices of from £300 to £600/m^3 could easily be obtained for veneer quality timber in 1986. The best cherry, with a uniform, honey-coloured wood sold for £1,700/m^3 in Germany in 1987, though butts with a variable colour and "green" rings were worth only £170/m^3. The average density of the wood at 15 per cent moisture content is about 630 kg/m^3.

PSEUDOTSUGA MENZIESII (Mirbel) Franco

Douglas fir

Origin and introduction

There are six species of *Pseudotsuga*: four in eastern Asia and two in western north America. Douglas fir has a very wide natural distribution and is a species of great commercial importance in its native north America. It occurs along the Pacific coast from northern British Columbia to northern California, the Rocky mountains and Mexico, and was introduced to Britain in 1827.

Climatic requirements

The species is moderately accommodating but tends to grow best in the wetter western parts of the country, and satisfactorily though more slowly in the lower rainfall areas of south-east and eastern England, provided there is sufficient soil moisture. It can be damaged by late spring frosts in low-lying places, though less severely than the spruces (Macdonald *et al.*, 1957). Douglas fir will not tolerate serious exposure, becoming badly deformed on exposed sites; it tends to be more prone to windthrow on unsuitable sites than many conifers. The tree requires reasonably sheltered conditions for healthy growth. Douglas fir is one of the least tolerant of all species to atmospheric pollution.

Site requirements

Douglas fir is usually regarded as a species for middle valley slopes and bottoms of at least moderate fertility. It needs a deep, well drained soil in order to develop a good root system. The soil can be clayey, if on a slope, and the tree will grow well on sandy soils. Adequate moisture and soil aeration are essential. Douglas fir is therefore unsuited to heavy waterlogged soils where rooting is restricted when the tree becomes very unstable; it is one of the least tolerant of all species grown in Britain to the anaerobic conditions resulting from flooding. It is also unsuited to calcareous soils and, like Sitka spruce, will not grow in competition with heather. On sites where it will thrive, especially in the wetter parts of the country, Douglas fir is one of the most desirable species as it is both productive and has a valuable timber.

Other silvicultural characteristics

Douglas fir is difficult to grow in mixtures with other species, especially broadleaves, because of its rapid rate of growth and its tendency to spread widely. Grass control is essential for successful establishment of young plants, but the species is more competitive with weeds than many other trees, and eventually casts a very heavy shade.

The tree is fast-growing, especially in terms of diameter in relation to height, and its stems have a tendency to sweep when young. It is sufficiently shade bearing to be useful for planting beneath well thinned canopies, especially for enriching scrub and neglected woodland, but not so shade bearing that it will form a lower storey to another species. Its litter decomposes readily on most sites. Branches tend to be very persistent, which results in knotty timber, unless stems are pruned. It is less susceptible to *Heterobasidion annosum* than many conifers.

Natural regeneration occurs patchily, but seldom on a sufficient scale for it to be a reliable means of crop replacement, in contrast to Sitka spruce in many areas.

In general, Douglas fir is a healthy and well adapted tree to many parts of Britain. The woolly aphid, *Adelges cooleyi*, which feeds in spring and early summer, causes yellowing and deformity of the needles and can severely check the growth of some provenances when the trees are young.

Flowering, seed production and nursery conditions

The tree flowers in April, seeds ripen, and are ready for collection in September, and are dispersed naturally from then until late March. (Cones should be collected when they are a light golden brown or yellow colour.) The earliest age at which the tree bears seeds is 30 to 35, but the best seed crops occur at intervals of four to six years, usually from the age of 50 or 60. This may be a greater age than normal commercial rotations. There are approximately 87,100 seeds/kg (range 65,100 - 100,100), of which 70,000 are normally viable. Seed collected in Britain is often infested with a seed wasp, *Megastigmus spermatrophus* (Forestry Commission, 1961), which can seriously reduce the quantity of sound seed (Kennedy, 1974). For nursery purposes, seed is usually stratified or pre-chilled for three or more weeks and sown in late March.

Figure 22. Douglas fir, *Pseudotsuga menziesii.*

Provenance

The best origins for use in Britain come from low elevations with high rainfall in the State of Washington, from the western foothills of the Cascade mountains, westward along the 46th parallel to the low coastal range and north as far as Forks in Clallam County (Lines, 1987).

Area and yield

Douglas fir has been planted on 47,000 ha, or 2 per cent of forest land in Britain (Locke, 1987). Mean yield classes range from 10 to 15 m^3/ha/year in different regions of Britain (Nicholls, 1981), and the maximum is 24. Rotations of maximum mean annual increment are similar to those for Sitka spruce.

Timber

Knot-free Douglas fir is one of the world's finest coniferous timbers, valued for veneering, joinery and decoration. It tends therefore to attract higher prices than most other conifers. Little clear joinery wood is produced in Britain, but most is suitable for carpentry and structural work, or for use in the round as transmission poles and fence posts. The timber dries with little degrade and is fairly stable, presenting no special difficulties in sawing and working, though it has a tendency to split when nailed. The wood is much less permeable to preservatives than Scots pine, so poles take longer to treat and tend to "bleed" in service. The average density of the wood at 15 per cent moisture content is about 530 kg/m^3. It is often sold as "Oregon pine".

QUERCUS L. — Oak

The genus *Quercus* contains about 600 species which are confined to the northern hemisphere: 27 are European species (Mabberley, 1990). Most occur in more southerly latitudes than Britain, in warm temperate and even tropical montane climates. By far the greatest concentration is in Mexico, which has about 125 endemic species and a further 75 which extend either north into the USA, or south towards Columbia. Only four oaks have any potential in Britain, of which two are native.

QUERCUS CERRIS L. — Turkey oak

The main virtues of Turkey oak are its resistance to exposure and adaptability to calcareous soils, even where they are shallow. It can be of value as a windbreak, especially in the south of England. Unfortunately its timber is much inferior to that of most hardwoods and this limits its potential uses considerably.

Turkey oak is indigenous to southern Europe and south-west Asia, and was introduced to Britain in 1735. The tree is superficially very similar to the two native species of oaks, with which it is easily confused. It grows to considerable sizes, and usually has excellent form. A good-looking oak should always be suspected of being a Turkey oak! *Q. cerris* grows faster than the *Q. petraea* and *Q. robur* and coppices well. It seeds regularly, and regeneration is common. Macdonald *et al.* (1957) considered that if the tree were more widespread it would speedily naturalise itself, as sycamore is doing.

Turkey oak is the alternate host of the knopper gall wasp, *Andricus quercuscalicis*, which became common in the 1980s causing concern about seed production and consequent regeneration of the pedunculate oak. It forms very characteristic convoluted galls on the acorns of *Quercus robur*, and has a complex life cycle involving the development of an asexual generation of the wasp, on pedunculate oaks, alternating with a sexual generation which develops in tiny galls on the male flowers of Turkey oak (Speight and Wainhouse, 1989). Knopper galls tend therefore to be much more common on *Q. robur* if there are Turkey oaks growing in the vicinity.

The timber is hard and brittle, and has an even wider band of sapwood than the native oaks. It seasons poorly, shrinks considerably on drying, and has a limited natural durability. It is suitable only for rough work such as shuttering and mining timber, and as firewood. Its average density at 15 per cent moisture content varies between 800 and 880 kg/m^3.

QUERCUS PETRAEA (Mattuschka) Liebl. Sessile oak
QUERCUS ROBUR L. Pedunculate oak

Origin

Detailed accounts of the two British oaks have been written by Jones (1959), Morris and Perring (1974) and Newbold and Goldsmith (1981) from which much of the following information is taken. Both species are native to Europe, including Britain and Ireland, and to western Asia. Within Britain, *Q. robur* is dominant mostly in the lowlands of the south, east and central parts of England, on heavy and especially basic soils. *Q. petraea* is characteristic on siliceous soils in the north and west of England, Wales, Scotland and Ireland. Their ranges overlap, on the damper acid sands, and planting over many centuries has obscured the differences in the natural ranges. Intermediates between the two species are quite common, and these are interpreted as hybrids (Clapham *et al.*, 1985).

Contrary to popular opinion, the British oaks are not "climax" species. Neither as seedlings, nor as adult trees are they really shade-bearing, but there are no large shade-bearers in Britain except for beech which has a more limited range. The widespread apparent "climax" role of the oaks is due to:

1) their ability to outlive most competitors, once established;
2) their ability to grow on a very wide range of soils;
3) good seed dispersal which enables them to spread.

Oaks have a considerable pioneer capacity: seed is dispersed widely by birds and mammals and their large reserves make seedlings competitive with most grasses. They can germinate quickly in the autumn, withstand summer droughts, and can persist in grass until the roots have developed enough to allow rapid shoot growth. Unlike most late successional trees the oaks are in no way dependent on shade from other trees. In natural conditions they regenerate by species alternation: for example, birch may replace fallen oak trees in northern oak woods, which are in turn replaced by other oaks, but because the oaks are so long lived, the forest remains predominantly of oak.

In environments where there are more shade-bearing "climax" species, such as in Illinois (Jokela and Sawtelle, 1985), there is widespread concern over the failures to obtain satisfactory regeneration of some species of oak (e.g. *Q. stellata, Q. imbricata* and *Q. marilandica*) which originally regenerated on former agricultural land. As they die of old age, they are replaced by the understorey of sugar maple. In this part of the USA the oaks behave as pioneers that gain early dominance in grassy areas. After one generation they are succeeded by more tolerant trees. *Q. petraea* behaves in the same way in the lower part of its altitudinal range where the more shade tolerant beech can be a serious competitor for dominance, especially on the more fertile sites. Beech is a far more stable climax species, being capable of regenerating under the shade of other trees, including self-replacement.

Climatic requirements

Within their ranges in Britain, the oaks are usually the dominant native trees. Sessile oak is said to be less frost tender though both species suffer badly from very late spring frosts; temperatures of -3°C will kill new foliage in the spring. They avoid earlier frosts by being among the last trees to flush, usually in mid to late April. Oak woodland is rare above 300 m altitude in Britain (Jones, 1959).

Site requirements

Though they will grow well and root deeply on heavy soils, oaks benefit from drainage (Evans and Fourt, 1981). The strong, deep rooting system allows them to evade droughts and makes them very windfirm. Shallow badly drained soils should be avoided. *Q. robur* also shows a preference for moist, heavy soils and it tolerates some waterlogging, while *Q. petraea* is intolerant of flooding, and is therefore found on better drained soils. Jones (1959) observed that *Q. robur* does better on more calcareous sites, rich in mineral nutrients while *Q. petraea* is more common on acid soils. Sites at low elevations with at least moderate fertility are required for acceptable rates of growth. Drought-prone and sandy soils, and soils with erratically variable water tables should be avoided as they give rise to a high proportion of trees with shaken timber.

The grasses *Deschampsia flexuosa* and *Holcus mollis* are both said to inhibit the development of seedlings, and the use of herbicides is necessary on such sites.

Other silvicultural characteristics

Many of the behavioral traits of *Q. robur* and *Q. petraea* are characteristic of trees in warmer climates. They come into leaf too late to make full use of the growing season, and even then are commonly damaged by frosts. The large fruits are readily killed by desiccation as well as frosts, and they germinate precociously. The ability of seedlings to shoot and die back repeatedly when growing in adverse conditions, while the taproot thickens and accumulates reserves until it is capable of producing a long, strong shoot in a single season is also a feature commonly found in warm climates; as is the ability to produce epicormics, which is an adaptation to fire. Coppicing ability is very good, especially in the south, and has probably been selected for by man to some extent through many generations of this treatment in Britain.

Both species are relatively intolerant to shade, but according to Jones (1959) *Q. petraea* is more shade-bearing. However, unlike most strong light-demanders, they are also rather slow growing. Early growth can be considerably hastened by the use of individual tree shelters.

Constant problems in producing high quality timber are caused by the growth of epicormic shoots and the economic losses associated with shakes in the timber, both of which are believed to be highly heritable (Kanowski *et al.*, in press).

Epicormic shoots appear when the space and light available to a tree is suddenly increased by thinning, and also as a response to restriction of the crown. The suppressed trees die from the top downwards, and as they do so new epicormics form at successively lower levels. Generally the better developed the crown, the smaller is the tendency to form epicormics. Jones (1959) recorded that when the ratio of diameter of crown to diameter of bole is increasing, there is little or no tendency to form epicormics, and the tendency varies with the rate of increase or decrease of the ratio. Wignall *et al.* (1985) found that thinning oaks in summer resulted in less epicormic production than spring thinning, and sessile oak is said to be less prone to developing them than pedunculate (Evans *et al.*, 1982). The conventional way to minimise epicormic production is to thin lightly and often. In continental Europe a lower storey of shade bearers, usually beech or hornbeam, is often grown with oak to shade the stems and hence suppress epicormic growth.

Shakes in the timber occur most commonly on drought-prone sites (Henman, 1984). Water stress is thought to "trigger" shakes in trees which have a predisposition to the problem. Savill (1986) found that earlywood vessels with a greater diameter than average provide the main predisposition. Vessel size has subsequently been found to be a very highly heritable characteristic of oak trees (Kanowski *et al.*, in press), so that it will theoretically be possible to select, or to breed trees with small vessels, though whether this will ever be a practical possibility is questionable. Fortunately, it has been found that trees in any population which come into leaf latest in any year, are also those with large earlywood vessels, and hence the most predisposed to shake (Savill and Mather, 1990). This simple way of recognising trees with a potential to shake makes it possible to remove them during early thinning operations, leaving shake-free individuals to grow to the end of the rotation. If trees are marked for removal in thinnings during the two to three week period of flushing, most of those with a predisposition to shake can be removed over three or four thinning cycles. The reason for the connection between flushing time and the possession of larger than average sized vessels is obscure, but it is possible that the same auxin (indole-3-acetic acid) is responsible for both late flushing and the production of large vessels.

Pedunculate oak has been used far more commonly than sessile in plantations, possibly because it seeds more often and the acorns are slightly easier to store. In some years it can be difficult or impossible to obtain *Q. petraea* seedlings from nurseries. The irregularity of seeding and the difficulty of storing acorns in a viable condition for more than a few months have always been serious problems to those concerned with planting. Attempts at preserving acorns for long periods on a large scale have come to nothing. Among those which have been reported, the earliest is probably by Ellis (1768) who found that cleaned and dried acorns enclosed in bees' wax remained viable for a year at least "the success of which, if properly followed, may in a few years put us in possession of the most rare and valuable seeds in a vegetating state from the remoter parts of the world, which in

Figure 23. A) Red oak, *Quercus rubra*, B) Sessile oak, *Quercus petraea*, C) Turkey oak, *Quercus cerris*, D) Pedunculate oak, *Quercus robur*.

time may answer the great end of the improvement and advancement of our trade with our American colonies."

Sessile oak is frequently reported to have a better form, tending to be straighter and taller, with a greater length of clear bole. The author of this book, at least, has yet to be convinced of this difference! Most of the older oak trees in Britain were grown as standards over coppice and in these conditions the trees are likely to be significantly shorter than those grown throughout their lives in high forest (Savill and Spilsbury, 1991). As Marshall (1803) observed, presumably about widely spaced trees: "Oaks which endure for ages, have generally short stems; throwing out, at six, eight, ten, or twelve feet high, large horizontal arms; thickly set with crooked branches....". Good comparisons of standards and high forest trees in reasonably close proximity can be seen in parts of Germany, such as Iphofen forest in Bavaria, where the differences in height at similar ages are very striking.

In Britain, more species of insects are associated with oaks than with any other trees, or with any other plant (Morris, 1974; Southwood, 1961). This accounts for the considerable value placed on oaks by everyone interested in nature conservation. By implication, it also means that, for those concerned with production, the oaks might suffer from some problems. Among the most serious pests are the caterpillars of *Tortrix viridana* and *Operophtera brumuta*, which cause serious, and even complete defoliation soon after the first leaves flush in spring: and the knopper gall wasp, described in the section on *Q. cerris*, causes some damage to acorn crops.

Young oaks are badly attacked by grey squirrels which strip the bark from the upper stem and branches. Unless control can be carried out effectively, it is probably not worth attempting to grow oak for timber.

Natural regeneration

Natural regeneration of oak is usually accomplished in Britain with difficulty, when it can be accomplished at all. Some writers (e.g. Rackham, 1980) believe that the lack of regeneration is a feature of the last 150 years and has been caused by the abandonment of the coppice system of management and increases in the numbers of small mammals, pheasants, squirrels, pigeons and deer which feed on acorns and browse on seedlings. In addition the introduction of the oak mildew in about 1907 may be responsible for killing shaded seedlings (Newbold and Goldsmith, 1981). Others, perhaps the majority, believe that oak regenerates as well today as it did in former times, though this has never been easy if quick establishment is considered important.

There are generally intervals of at least two, but often up to ten years between heavy acorn crops. In the west, germination occurs freely almost anywhere, but is best under litter which has a very suitable microclimate and also offers some protection from predation and frost damage. Temperatures of -6°C for a few hours can kill most acorns (Jones, 1959). In the drier parts of south-east England, acorns on the surface fail to germinate because they desiccate (Watt, 1919). The dormancy of acorns is never very deep, and in suitable conditions some germinate immediately on falling, but the epicotyl always remains dormant until it has been chilled over a winter. Dry autumns can be very damaging in that they cause the death of tap roots which emerge on the surface. If summer droughts occur, the acorns which germinate late usually die quickly. Burying acorns by scarifying the soil often results in earlier germination and much more prolific regeneration of oak, in just the same way as has been described for beech. It is often an advantage therefore to create conditions which encourage autumn germination of the radicle, so that the seedlings can begin growth early in the spring. High winter water tables also result in excessive mortality of acorns and seedlings, especially of *Q. petraea*. This is an important reason for retaining a canopy during the early establishment phase on heavy soils because water tables usually rise on clear felled sites. Maximum height growth of first year seedlings occurs in about 30 per cent of full light, (equivalent to open canopied woodland or small clearings), but at least 50 per cent is needed in later years (Jarvis, 1964).

Young oaks are well adapted to survive some browsing by wood mice (*Apodemus sylvaticus*) and bank voles (*Clethrionomys glareolus*), but they do not survive complete defoliation for many years. Defoliation of seedlings by the winter moth (*Operophtera brumuta*) and the green oak tortrix (*Tortrix viridana*) and many other lepidopteran caterpillars which come down from the crowns of mature oaks before pupation is one of the main reasons why oaks will not regenerate well under parent trees. Established trees are able to withstand repeated attacks by producing chemical defences (tannins), and because they produce several flushes of new leaves each year. For maximum growth and reproduction most caterpillars must complete feeding before any appreciable amounts of condensed tannin is laid down in the leaves. Tannin contents vary very widely from tree to tree.

Where natural regeneration is managed successfully, it is therefore necessary to open the canopy very rapidly once seedlings are established on the ground if they are to survive attacks by defoliators: hence the use of the uniform shelterwood system with these species is essential. Attempts to manage oak by selection or group systems (except by using very large groups) are doomed to failure, and the concept of an intimately mixed, all-age climax oak woodland is impossible.

Flowering, seed production and nursery conditions

Both species flower in May, acorns ripen between September and November, when they are ready for collection. The earliest age at which the oaks bear seed is 40 to 50 years, but the best crops are usually after the age of 80 years. Good seed years occur on average every three to five years, though exceptionally they can be at ten yearly intervals. *Q. robur* is a better seed producer than *Q. petraea*, possibly because it is more common in the lower-lying, fertile and warmer parts of the country. In *Q. petraea* there are commonly 316 acorns/kg (range 130 - 649), and in *Q. robur* the acorns are slightly heavier, at 273/kg (range 110 - 495), in both species 80 per cent of them will normally germinate. For nursery purposes the acorns should be stored in a cool well ventilated place, and from January onwards. If signs of shrivelling are observed, they should be sprinkled with water until sowing in late March. They can also be sown in the autumn in well drained soils, and if bird predation is unlikely to be severe: this can be provided by covering the seedbeds with 7 to 10 cm of extra soil, which is removed in March. Apparently damaged tips to radicles do not affect germination (Aldhous, 1972).

Provenance

There has been very little work on this subject in Britain or anywhere else, though provenance research began in 1987. Evans (1981) states that oak from the Forest of Dean grows well and is of above average form. Carpathian (Romanian) oak is excellent, and New Forest and Scandinavia origins are poorer than average.

Area and yield

Oaks occupy 190,000 ha (9 per cent of forest land) in Britain, making them together the third most common species after Sitka spruce and Scots pine. Mean yield classes are low, averaging only 3 to 5 m^3/ha/year (Nicholls, 1981), with a maximum of 8. Rotations for sawn wood or veneer are long, commonly 120 years or more. The best veneer trees in the Spessart region of Germany are grown on 400 year rotations. For economic reasons, oak is often grown with a productive conifer, such as Norway spruce or European larch, which will provide some financial return to the owner after 50 or 60 years, leaving a pure crop of oak for the remainder of the rotation.

Timber

Good oak timber is used for furniture, panelling, high class joinery, and veneers. Its attributes of a decorative appearance, natural durability and great strength make it very valuable for outdoor work such as fencing, gates, and mining timber in the lower grades. The average density of the wood at 15 per cent moisture content is about 720 kg/m^3. The valuable veneer oak grown in the Spessart region of Germany sold for £1,000 to £1,400/m^3 in 1987 and must have annual rings little more than one millimetre wide. Generally light coloured oak is rare and, because of this, it fetches the highest prices.

Oak timber can be remarkably difficult to sell profitably. Its tendency to shake, and to grow large, heavy branches, and its relative difficulty in working all present problems in the modern world of bulk, standardised material. Even its natural durability is less of an asset today, when most timbers can be treated with preservatives quite easily. Only the best trees pay for themselves, and these can be very profitable. Oak which has been grown very slowly, with annual rings less than one millimetre wide, is weak and brittle, because it is composed almost entirely of springwood, made up of large diameter vessels. It is very light when dry. The later formed summer wood gives oak its strength. Hence, the faster oak is grown, the stronger it is. This is true of other markedly ring porous species, such as sweet chestnut, as well.

It has long been known that oaks should not be felled when leaves are on the trees. Theophrastus (c. 200 BC) stated "if it is cut at the time of peeling, it rots almost more quickly than at any other time....What is cut after the ripening of the fruit remains untouched by worms...." This is because the sapwood contains sugars and other carbohydrates during the growing season which make it much more susceptible to attacks by fungi and insects.

QUERCUS RUBRA L. — Red oak

Origin

Red oak has a very wide distribution in the eastern USA and the extreme south-east of Canada. It is the most northerly of the eastern American oaks and grows from valley bottoms up to the lower- and mid-slopes of hills and mountains. It is a major species in the hardwood forests of eastern North America and was introduced to Britain in the early 1700s.

Silviculture

The species is hardy in Britain, but like the two native oaks is occasionally damaged by late spring frosts. It grows rapidly even on the poorest acid soils, though not on peats, and does best on acid sandy loams. It will not grow well on calcareous sites. The French believe it has a place on dry, acid sites for which there is no productive broadleaved species at present. It could, for example, be planted in some of the drier parts of lowland Britain on sites too poor for sweet chestnut, and where atmospheric pollution is too high for conifers, but where reasonably rapid growth is required. Like *Q. robur* and *Q. petraea*, the timber of this species has a reputation for being shaken. Assuming the causes of shake are the same, the use of the species on well drained and drought-prone sites is likely to lead to timber which will very commonly be shaken. On such sites, red oak should probably be regarded as an amenity species, and certainly not as a species for producing saw timber.

Red oak is more shade-bearing than the native oaks and will form a good understorey in pine stands. It coppices well and is known for its crimson autumn colour for which it is, perhaps, most often planted. Unlike the native oaks, epicormic branches are not a serious problem, but it has a tendency to fork rather badly, and so must be cleaned and pruned early. Frequent heavy thinnings are also considered necessary in France. Macdonald *et al.* (1957) stated that it is not a long-lived tree, and will not grow to the dimensions of the native species.

Flowering and seed production

The tree flowers in May; seeds ripen and are ready for collection in September and October. The earliest age at which the tree bears seed is 30 to 40 years. The best seed crops are produced at intervals of two to four years 20 to 40 years later, mostly in the south of England, but even there they are not plentiful. It is probably because of the failure to produce viable seed in adequate quantities that it has remained a minor species in the UK. There are about 280 seeds/kg (range 165 - 564), of which 80 per cent will normally germinate.

Area and yield

There are said to be about 700 ha of *Q. rubra* in Britain, and it is widely planted in many parts of western continental Europe. In France, for example where it is an important tree, there is a large programme of selection and breeding.

Levels of production are usually much higher than with the native oaks. In the Netherlands they range from yield classes of 3 to 9 m^3/ha/year (Bastide and Faber, 1972), up to 9.4 m^3/ha/year in Belgium (Laurent *et al.*, 1988), and in Britain similar levels have also been found.

Timber

In Britain, the timber is traditionally thought of as being inferior to that of the native oaks for furniture and decorative work because its colour and texture is less attractive. It is nevertheless in demand in France for furniture-making and is also used for flooring and interior joinery. Untreated timber is unsuitable for exterior work because of its lack of natural durability, though it can easily be impregnated with preservatives. It is at least as strong as the native oaks (FPRL, 1964). The average density of the wood at 15 per cent moisture content is about 790 kg/m^3.

ROBINIA PSEUDOACACIA L. Locust, False acacia

There are three or four north American species of *Robinia* which are leguminous, nodulated, nitrogen-fixing species. This attribute, which is rare among trees that will grow in Britain, often arouses interest because of its potential use as a nurse or pioneer species in "stressed" environments. In the UK, *R. pseudoacacia* is a low volume producer, and has a very bad form, and so is unlikely ever to be grown as a timber-producing species in its own right.

Origin and introduction

The species is native to the eastern and mid-western USA between latitudes 35° and 43°N, where there are two separate populations each side of the Mississippi valley. It was introduced to Britain before 1640.

Site requirements

Robinia grows reasonably well on dry and infertile sandy soils. Its possible value as a nurse on such sites, especially those poor in nitrogen (such as strip-mined areas) is the species' main asset in Great Britain.

Silviculture

The tree is seldom planted in Britain today because of the nuisance caused by its prolific suckers, and the unpleasantness of its thorns (Macdonald *et al*., 1957). In its native habitat, *Robinia* is an important colonising species in areas of disturbed forest. It is not very long-lived, but rapidly regenerates from seed, stumps and root suckers. Root suckers grow particularly well from exposed roots following soil disturbance, so that the species has a value in the control of erosion on slopes and in gullies. In the USA young trees, particularly in the diameter at breast height range of 2 to 12 cm, are very badly attacked by a borer, *Megacyllene robiniae*, especially on nutrient-poor sites, to the extent that in 1913 planting of the species was virtually given up. Thirty years later it returned to use because its tolerance to acid soils and nitrogen fixing properties made it ideal for planting on strip-mined sites, particularly in the Appalachians.

Its broad site tolerance and silvicultural properties have resulted in a wide acceptance in continental Europe. In Hungary it is now a species of major economic importance, as well as in Romania and France, where it is valued both for timber and nectar production by beekeepers, and also as a forage species (Keresztesi, 1983). *Robinia* is now rivalling poplar as the second most widely planted broadleaved species in the world, after the eucalypts. Selection and breeding programmes are being extensively practised in eastern Europe.

Seed production and nursery conditions

There are about 53,000 seeds/kg of which 70 per cent will normally germinate. For nursery purposes seed pods have to be picked from the trees. Like those of many legumes, the seeds of *Robinia* have very impenetrable coats which can delay germination. A satisfactory method of pre-treatment is to put the dry seed into a container with five times its volume of freshly boiled water and leave it to cool. Aldhous (1972) stated that this is more reliable than the alternative of soaking in cold water for a week. It should be sown at the end of March or early April.

Timber

The timber is naturally extremely tough (similar to ash), very durable, with an unusually narrow (c. 0.5 cm) sapwood. It is light (about 740 kg/m^3 at 15 per cent moisture content), hard and tough which makes it very suitable for outdoor work. The wood is usually straight grained and fairly coarse textured. It has a strong tendency to warp. In the nineteenth century, it was used for ship building in the United States. The wood is naturally yellowish green. It has excellent properties for steam bending, and high pressure steaming can change the colour to golden-yellow, yellowish brown, light brown or dark brown. The larger logs can be used for veneers.

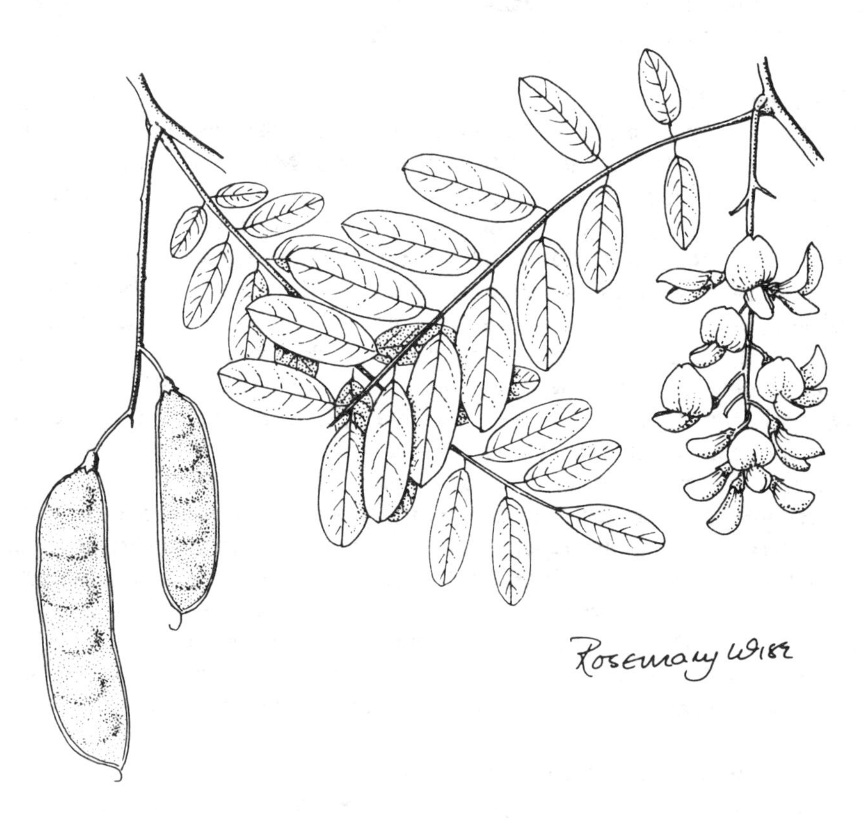

Figure 24. False acacia, *Robinia pseudoacacia.*

SALIX L. Willows

Species

About 400 species of willow occur worldwide, though most are in the north temperate and arctic regions. It is generally agreed that 18 species and 27 interspecific hybrids are native to Great Britain and Ireland. The tendency to hybridise, especially among sallows, make members of the genus very difficult to identify. The tree willows native to Britain are *Salix pentandra, S. caprea, S. fragilis*, and *S. alba*. None of them is of any importance as a productive forest tree, but numerous species and varieties of willows are commonly planted as ornamentals from cuttings, by which all members of this genus are normally propagated. One of the most popular is the weeping willow, *S.* x *sepulcralis.*

Site requirements

Species of willow are often the first woody plants to become established in wet habitats, and they can be valuable for stabilising the banks of waterways. They tolerate the low soil oxygen levels around roots in wet places by producing numerous long adventitious roots, to increase the surface area for absorption.

Most tree willows require deep, rich soils for reasonable growth, but they are not very long-lived.

Silviculture

Most of the pollarded willows that line the banks of English lowland rivers are *S. alba*. A variety of it (var. *caerulea*), is the well-known cricket bat willow of which about 2,000 ha are grown in Britain on two to five year rotations, mostly in Essex (Evans, 1984). Like poplars, they do best at very wide spacings on stream and river banks where there is moving (aerated) water. An account of cricket bat willow cultivation has been written by Warren-Wren (1965). A potential threat to growing satisfactory trees for cricket bats is the "watermark" disease, caused by the bacterium *Erwinia salicis*. This causes wilting and dieback, and the wood becomes brittle and unsuitable for the manufacture of bats (Preece 1977; Gibbs, 1987). It is controlled by propagating from disease-free parent trees and by the prompt destruction of infected trees.

Willows also have potential, though little used values in silviculture in being planted as "sacrificial" trees in places where deer browsing and rubbing is a problem. Deer prefer willows to most other trees. It has been suggested that this might be because of the salicin content of the bark which, like aspirin (from which it originated), can relieve pain and irritation, especially when the felt on antlers is being shed. Planting them for this purpose can relieve the pressure on more commercially valuable species.

There is a great deal of recent interest in growing willows, and poplars, on five or six year rotations as short rotation biomass crops for energy and industrial uses (Hummel *et al.*, 1988). The important species for this are *S. viminalis* and

S. aquatica gigantea, though the latter has become very susceptible to the fungus *Melampsora*. It is becoming increasingly obvious that no single clone can be used very extensively, or for very long periods, because of risks of diseases.

Willows are hosts to a very large number of herbivorous insects. Species of willow, especially *S. fragilis* and *S. alba*, are the woody winter hosts of a very damaging aphid of carrots, *Cavariella aegopodii* (Jones and Jones, 1974).

Timber

The wood is used for purposes where light weight, easy working and resistance to damage by impact are required: hence its value for cricket bats, and at one time for the wooden parts of artificial limbs. It weighs about 450 kg/m^3 at 15 per cent moisture content, though cricket bat willow is lighter, at 340 to 420 kg/m^3.

Before the days of plastics and other similar materials, willows were enormously important in the economy of the country for the production of basket work for numerous purposes: hurdles, screens, and containers of all kinds. The leaves were used for animal fodder at one time.

SEQUOIA SEMPERVIRENS (D. Don) Endl.

Coast redwood

Origin and introduction

The coast redwood occurs naturally in a narrow part of the "fog belt" along the coast, from the extreme south of Oregon to south of Monterey in California. The tallest tree in the world is a member of this species (112.4 m tall). It was introduced to Britain in the 1840s, and is characterised by a very thick, soft, fibrous red bark which can be punched without causing injury to the hand.

Silviculture

Macdonald *et al.* (1957) stated that the best trees tend to be found in the moister, south-western parts of Britain on particularly favourable sites, though impressive specimens can be found in most lowland parts of the country. It is reasonably shade bearing, easily damaged by frosts, and consequently established with difficulty in the open. It must have adequate shelter from the wind and will not grow well on very acid soils. It cannot tolerate serious atmospheric pollution.

The main interest in this species is its potential for rapid growth and very high volume production, but only on really good sites. According to Edlin (1966), it grows much faster than Sitka spruce on the best sites. One stand at Leighton in Powys had a standing volume of 2,152 m^3/ha at age 94, and was said to have the highest volume standing on the ground of any stand in Great Britain, by Macdonald *et al.* (1957). As Edlin (1966) has said, most British foresters exclaim that it is "a tree of great promise", but that is as far as it ever goes. One reason why it is seldom planted is that it is very difficult to raise in any quantities from seed: often well under 10 per cent is viable, and then seedlings can be severely injured by frosts. Plants can relatively easily be raised from cuttings and by micro-propagation however.

A rare feature among conifers is the ability to produce coppice shoots, but redwoods do it with great vigour, and on suitable sites such as at Longleat in Wiltshire, the crop can be renewed by this means very satisfactorily. It is also said to produce root suckers (Mabberley, 1990).

Timber

The wood turns a maroon colour soon after felling (hence its name), and in its native range the trees have supported a large timber industry. The timber has the reputation of being strong and naturally durable. Its main drawback in use is said to be its softness (Gale, 1962). The wood is light, with an average density of about 420 kg/m^3 at 15 per cent moisture content. It is non-resinous and stable in service. It is used in America in house building, and because of its durability, for such things as shingles, cooling towers, silos, greenhouses and farm buildings.

SEQUOIADENDRON GIGANTEUM (Lindl.) Buchholz
Giant sequoia, Wellingtonia

Origin and introduction

The giant sequoia is native to the middle elevation mixed coniferous forests of the central and southern Sierra Nevada in California. The species is famous for the huge dimensions and great age to which it grows: in 1931, the biggest, the "General Sherman" tree, had a height to the top of the trunk of 83 m, and a volume of 1,406 m^3. It has not apparently been measured reliably since then. It was introduced to Britain in 1853, having been discovered in 1850.

Full accounts of the species and its ecology have been written by Hartesveldt *et al.* (1975) and Harvey *et al.* (1980). It is described as a fire subclimax species which is intolerant to shade, but relatively resistant to fires which set successions back to an earlier stage.

Figure 25. A) Coast redwood, *Sequoia sempervirens*;
B) Giant sequoia, *Sequoiadendron giganteum.*

Silviculture
The giant sequoia has never gone beyond the trial plot stage in Britain, though it is quite widely planted in parks and gardens. Macdonald *et al.* (1957) state that on very limited evidence it is less badly affected by exposure and frost than the coast redwood, and will grow on most soils except wet acid peats, and in most parts of the country. Losses can be high after planting. The tree grows very slowly for the first few years, and weed control must be thorough. This species is less tolerant of shade than the coast redwood but is slightly more tolerant to atmospheric pollution.

Timber
The wood is even less dense than that of *Sequoia*, at about 340 kg/m^3, but is extremely durable. It is said not to be very useful commercially (Patterson, 1988), except for roof shingles.

SORBUS ARIA (L.) Crantz **Whitebeam**

Whitebeam is a small tree which occasionally grows to 20 m tall. It is a light-demanding, early coloniser, which is usually found on chalk and limestone, and also on sandstone hills from Kent and Hertfordshire to Dorset and the Wye valley. Like the next species, rowan, it has considerable decorative value and is often planted for this purpose.

Whitebeam produces showy white flowers in the spring or early summer, greyish-green leaves, and red berries in the autumn. Propagation in the nursery normally requires the berries to be macerated in order to separate out the seeds, which should then be stratified and sown the following spring. If the seed is not separated, it may remain dormant for an additional year (Aldhous, 1972). There are about 1,500 seeds/kg of fruit.

SORBUS AUCUPARIA L. **Rowan, mountain ash**

Rowan is widely known as a small ornamental tree which produces numerous, attractive small white flowers in May and very handsome scarlet fruits in the autumn. It is native to British woods, scrub and mountain areas and will thrive at elevations up to almost 1,000 m on sites where few other broadleaved trees will grow at all. It is found mainly on lighter soils in the north and east, and is rare or absent on clays and soft limestones. Today, it is best known for its ornamental value both in wild upland parts of Britain, and in the form of numerous named cultivars in gardens, where it has the additional virtue that it does not grow too big. Rowan is a strong light-demander and it coppices well.

An aphid, *Rhopalosiphum insertum*, migrates from rowans in summer to the roots of oats, other grasses and reeds, and can cause some economic damage.

The tree first fruits at age ten, and is a very prolific producer from 15, either every year or every other year. The fruits ripen between July and August and seed treatment for propagation in the nursery is the same as for whitebeam. There are about 5,000 seeds/kg of fruit.

The wood is not durable but is sometimes used for turnery, furniture and engraving. It is very similar to that of apple.

SORBUS TORMINALIS (L.) Crantz **Wild service tree**

Interest in the wild service tree arises from its rarity, the challenge it presents in being grown to large sizes and the potentially valuable timber. Its yellow autumn colour is also most spectacular. The leaves are very similar to those of maples, except that they are alternate rather than opposite, and it is frequently mistaken for a maple.

Origin

The wild service tree is found over much of northern Europe, the Mediterranean region and north Africa. It is native, to England and Wales, but has a patchy, local distribution from Cumbria and Lincoln southwards. The presence of the species is usually regarded as an indicator of primary woodland because regeneration from seed is very rare.

It will occasionally grow up to 25 m tall and about 80 cm in diameter, but is more commonly a much smaller tree, or even a large shrub.

Site requirements

It usually grows on clays, similar to those favoured by wild cherry, but also on limestone (Clapham *et al.*, 1985). In Lower Saxony the species is well adapted to dry, calcareous, but nutrient-rich sites as a timber tree, and it is particularly common near coasts (Meyer, 1980).

Other silvicultural characteristics

The tree can be badly damaged by deer and rabbits. Schmeling (1981) commented that, as a timber producer, it is not really suited to high forest conditions because its slow height growth and light-demanding nature not only prevent it keeping up with faster-growing species, but also cause it to become suppressed and eventually to die. It probably does best as a standard in systems which are coppiced every 25 to 40 years, or in the open. The decline in coppicing over the last 100 years has probably contributed to the comparative rarity of the species.

The wild service tree coppices well and older individuals also produce masses of very shade tolerant root suckers for a distance of 3 to 8 m around the stem, especially after the parent tree has been felled. These grow fast for 30 years or so, but unless they are then freed, they become suppressed and may die.

One reason the species has been so little used is that it is difficult to grow from seed, both in the nursery and in natural regeneration. Though it fruits every year or two, germination percentages are very low. Newly germinated seedlings need full overhead light, unlike suckers. Suckers are slow to grow to usable plants in the nursery, because they tend to have such inadequate root systems when collected.

Timber and uses

The wood is very similar in most respects to that of apple. It is valued for veneer, and making musical and, formerly, measuring instruments. It has a decorative grain and colour, and in continental Europe can command very high prices. Because of this, work is being undertaken in France (Lanier *et al.*, 1990) to determine the relationships between wood quality and site, and the silvicultural methods required to encourage its growth.

White flowers appear in the spring, and Cobbett (1829) described the fruit as: "a thing between a sloe and a haw. It is totally unfit to be eaten." The leaves and bark were reputed at one time to have medicinal properties. These are described by Evelyn (1678).

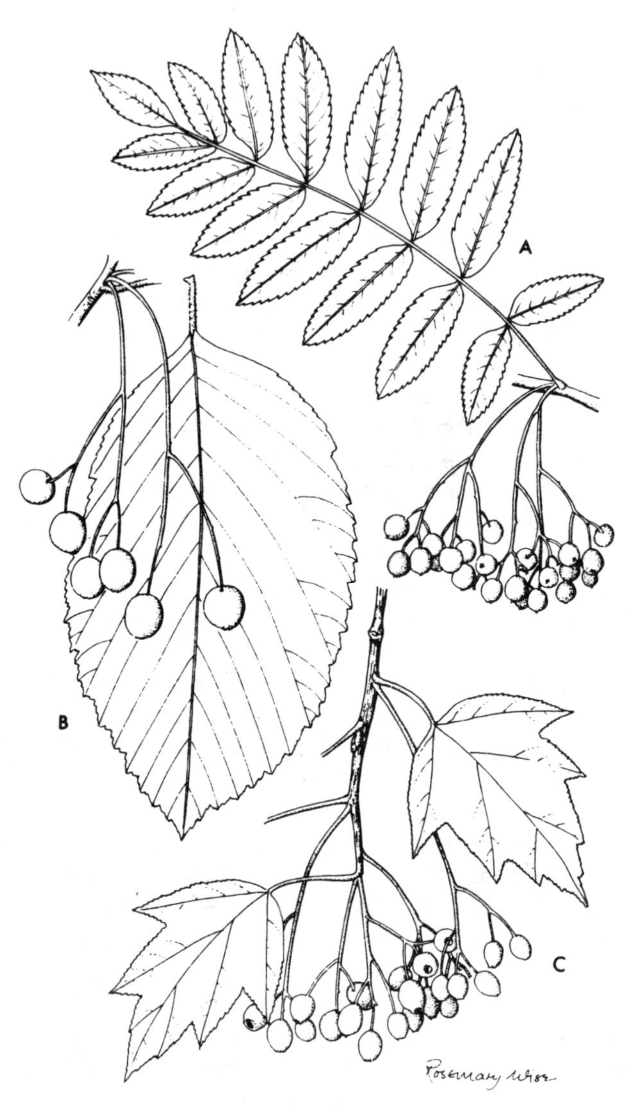

Figure 26. A) Rowan, *Sorbus aucuparia*; B) Whitebeam, *Sorbus aria*; C) Wild service tree, *Sorbus torminalis*.

TAXUS BACCATA L. Yew

Origin

There are seven species of yews in the northern hemisphere. *T. baccata* is native to Europe, including the British Isles, the Atlas mountains, and from Asia Minor to Iran.

Within Britain it is native to the chalk downs in the south of England, limestones in the north, and oakwoods on other soils. It has been very widely planted as an ornamental and in churchyards for many centuries. Several varieties exist. The tree has a reputation for living to a greater age than almost any other species which will grow in the United Kingdom.

Site requirements

Yew will grow on most soils so long as they are not waterlogged. The species sometimes forms pure woods in sheltered places on chalk in the south-east of England and on limestone in the north-west.

Figure 27. Yew, *Taxus baccata*.

Silvicultural characteristics and timber

Yew is one of the most shade tolerant of all British trees. It will withstand smoke and salty winds. The trees grow exceedingly slowly, which is why they are never planted as commercial forest trees. Individuals will grow to about 20 m tall. The wood of yew, at about 670 kg/m^3, is among the densest of all conifers. It was once used for making bows and knife handles.

Flowering, seed production and nursery conditions

Yew is dioecious. It flowers in the spring, producing very small catkins. The fruit is a hard seed, partly embedded in a pulpy, conspicuous and bright red, berry-like aril. Seed should be collected from the female trees in November, the arils removed by crushing, and the seed stratified until the March, 16 months later, before sowing (Aldhous, 1972). There are about 4,400 seeds per kg. The foliage and seeds of yew contain alkaloids which are very poisonous, though the aril itself is harmless, and can be eaten quite safely.

THUJA PLICATA Donn ex D. Don

Western red cedar

Origin and introduction

There are five species of *Thuja* in eastern Asia and north America. Western red cedar is native to north-west America, from Alaska to California and east Idaho. It extends from sea level to 1,800 m, though above 1,500 m it is a low shrub. *Thuja* is important in Douglas fir forests on the banks of water courses and in marshy valley bottoms. It was introduced to Britain in 1853.

Climatic requirements

In its natural range, western red cedar is restricted to areas with abundant rainfall or snow, high humidity and cool summers. It can withstand very low winter temperatures and is fairly resistant to late spring frosts. *Thuja* does not tolerate exposure well but is nevertheless fairly windfirm. In the UK it does best on relatively sheltered sites, such as the lower slopes of valleys, in regions with at least 700 mm rainfall, especially at low elevations in the west, but it grows perfectly well where precipitation is lower. It should not be planted at elevations much greater than 200 m, and lower in Scotland and the north of England.

Site requirements

Western red cedar is at its most competitive on the heavier-textured lowland soils in southern England, especially on calcareous clays. It requires at least moderate fertility and soils which are not too acid. It will succeed on wet and even rather dry shallow soils and is one of the best conifers for calcareous downlands. Deep, loose and infertile sandy soils should be avoided.

Other silvicultural characteristics

Western red cedar is a markedly shade-bearing species which will grow faster than other shade-bearers on heavy lowland soils. It is a particularly valuable conifer to grow in mixtures with broadleaved species as the narrow crown does not interfere with neighbours. It will keep pace with several species, such as the oaks, beech, ash and cherry without outgrowing or suppressing them, and will live with faster growing trees. *Thuja* is useful for enriching and underplanting and is also valuable in pure stands. Natural regeneration is prolific in some areas.

Western red cedar was difficult to grow at one time because of attacks by *Didymascella thujina* in the nursery (Forestry Commission, 1967) but this disease can now be treated with fungicides. It is more susceptible than most conifers to attack by *Heterobasidion annosum* and other fungi which cause butt rots. Often the bottom 2 m of a stem may contain rot. Affected stems are frequently swollen up to the point at which the rot ends. Small dead knots commonly rot, and therefore early pruning is advisable; since the foliage is in demand by florists, this can often be done very profitably at an early age. *Thuja* is less vulnerable to browsing by deer than many species, but it is occasionally attacked by grey squirrels.

Flowering, seed production and nursery conditions

The tree flowers from late March to early April and seeds ripen in September. The earliest age at which the tree bears seeds is 20 to 25 years, but the best seed crops are produced between 40 and 60. *Thuja* is normally a profuse seed bearer every two or three years. There are approximately 912,700 seeds/kg (range 447,500 - 1,307,100), of which 500,000 are normally viable. Cones should be collected as soon as they change from bright green to yellow, and the tips of the seed wings are visible and a light brown colour. This is normally in September. Seed is normally sown between late February and mid March, and needs no special treatment (Aldhous, 1972).

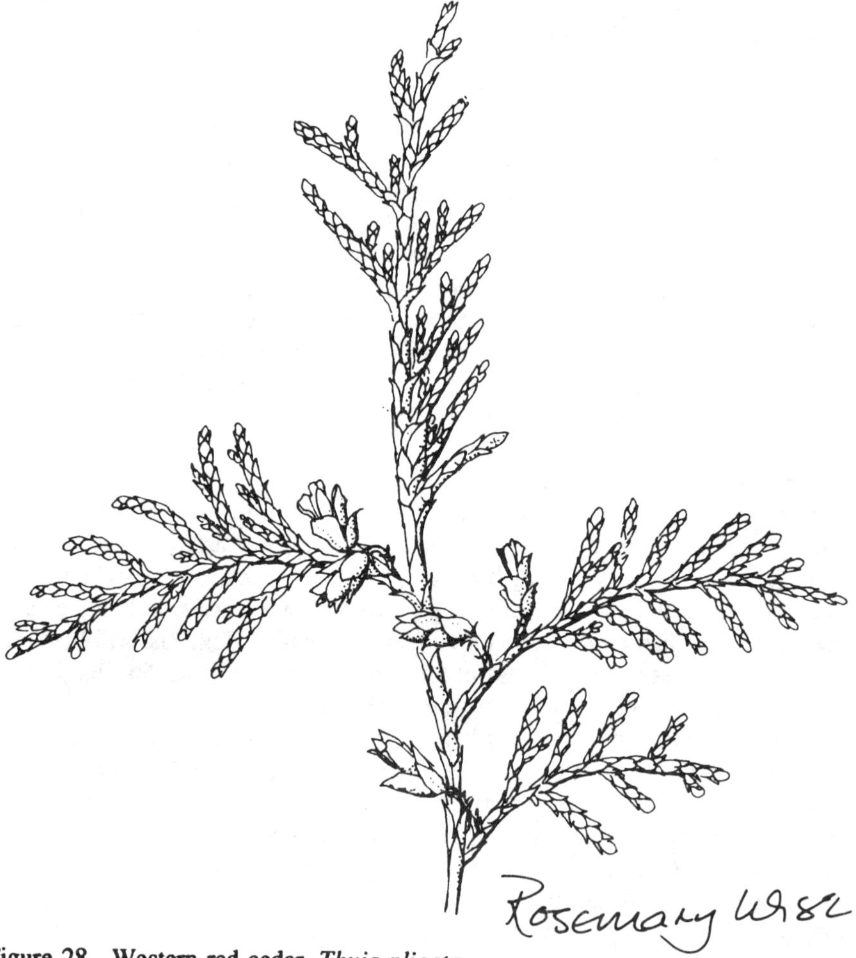

Figure 28. Western red cedar, *Thuja plicata*.

Provenance
Recommended origins for the UK are from the north slopes of the Olympic Mountains in Washington, and some Vancouver Island sources show promise (Lines, 1987). The stem tends to fork and the lower bole is often fluted with some seed sources, though excellent form can be found.

Yield
Yield classes range from 12 to 24 m^3/ha/year, with ages of maximum mean annual increment between 60 and 70 years.

Timber
The timber, in its native North America where it is sometimes known as arborvitae, is particularly noted for its natural durability. In Britain, the relatively fast growth (average density at 15 per cent moisture content is about 390 kg/m^3), and short rotations, makes it less durable, because there is often a much greater proportion of sapwood. The market for sawlogs may be high, but because of the red colour of the wood and its chemical constituents, small dimensioned *Thuja* is not acceptable for pulp and most other wood products, including chipwood. Young stems make ideal fencing rails which can be used without preservative treatment. There is a limited but lucrative market for poles (rugby posts, flag poles, etc), if they are grown slowly enough. It is also used for making garden sheds, beehives, and can easily be sold for fencing and allied uses. At least in North America, it produces the traditional shingles for roofing houses.

TILIA CORDATA Miller — Small-leaved lime
TILIA PLATYPHYLLOS Scop. — Large-leaved lime

There are 45 species of *Tilia* in north temperate regions, including five in Europe. The two native limes, like the oaks, tend to hybridise where their distributions overlap, as in the Derbyshire Dales and the Wye Valley.

Tilia cordata is native to England as far north as the Lake District, to Wales, and to continental Europe from northern Spain to Norway and eastwards to beyond the Urals. It is much the commoner of the two limes, and was the last major tree to reach Britain after the ice age.

Though the status of *T. platyphyllos* in Britain has been in doubt for a long time, Pigott (1988) states that on the basis of pollen analysis it is definitely a native tree. In continental Europe it is a more southerly species than *T. cordata*, extending from a single location in southern Sweden to central Spain, northern Greece and southern Italy eastwards to Poland, and the western Ukraine. In Britain it is found in more southerly regions than *T. cordata*. A third species, The common lime (*T.* x *vulgaris* Hayne), is a fertile hybrid between the two and is the most widely planted lime in gardens and towns.

Until Pigott's (1988) review of the ecology and silviculture of the species, remarkably little had ever been written on the limes. Much of the information below is from this.

Climatic requirements

Both species are essentially trees of the lowlands and lower ranges of hills.

Site requirements

The distribution of *T. cordata* in England and Wales is patchy, but the widely-held belief that it is a tree of limestone is incorrect. It will grow on podzols, brown earths and calcareous soils. In southern England it thrives on stagnogleys (soils which are waterlogged in winter and hard in summer), and on these sites it can compete successfully with oak. Only in the west and north, where the rainfall exceeds about 850 mm, does it move onto limestone as well as acid soils. Small-leaved lime grows in woods with oak and hornbeam. In prehistoric times it is believed to have been much more common than it is today; its decline is probably due to its selective removal by man.

T. platyphyllos is a tree of rendzinas derived from limestone, or basic igneous rocks, being found mostly with beech, ash, sycamore and yew.

Other silvicultural characteristics

T. cordata has a long history of being used in forestry, but only as a coppice species because it can produce long, and above all straight poles. The stools of lime show great longevity and appear to be almost indestructible.

Soils under lime tend to be much richer in nitrogen and phosphorus, and have higher populations of earthworms than under many other broadleaved species. This may partly be explained by its deep-rooting ability, but also the lime aphid, *Eucallipterus tiliae*, may cause an enrichment of nitrogen (Mabberley, 1990). This arises because they produce large amounts of honeydew and are said to deposit up to 1 kg of sugars/m^2 each year. These possibly stimulate the growth of nitrogen-fixing bacteria around the trees, and thus enhance nitrogen availability at the cost of some carbohydrate taken from the phloem by the aphids. Lime has often been planted for its soil-improving litter. Both species are very shade tolerant and can be grown as understoreys to oak and other species.

The limes are remarkably free of many diseases to which other trees are prone and are not damaged by squirrels.

On good sites, limes will grow at something like 8 m^3/ha/year. At present, most seed used in Britain is imported from Poland, and is believed to result in trees of a most undesirable form.

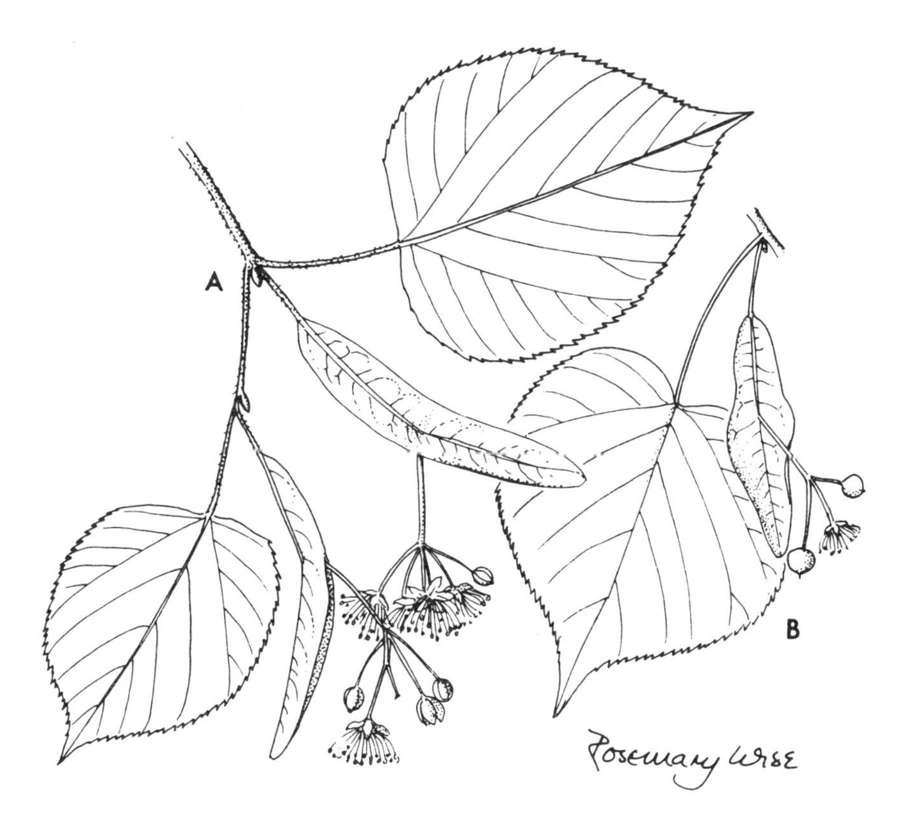

Figure 29. A) Large-leaved lime, *Tilia platyphyllos*;
B) Small-leaved lime, *Tilia cordata*.

Limes are attractive trees, and favourites for planting along roads in towns, presumably because of their sweet-smelling flowers and the fact that they can be pollarded and heavily pruned without coming to too much harm. *T.* x *vulgaris* and *T. platyphyllos* are particularly suitable as street trees because of their narrow-crowned habits, while *T. cordata* has too wide a crown. The lime aphids can be a nuisance on cars and pavements.

Flowering seed production and nursery conditions

The limes are among very few British tree species which are insect, rather than wind pollinated, and they, with sweet chestnuts, are among the last trees to flower in any year, usually in June or July. The flowers are sweet-scented and pale whitish-green. Contrary to popular belief both species produce viable seed often in southern Britain, but especially *T. platyphyllos.* In the north seed is usually sterile because the climate is too cold for the pollen tube to grow, and consequent fertilisation. Natural regeneration very rarely persists because it is so palatable to browsing animals, such as bank voles.

The trees first produce seed between 20 and 30 years of age, and every second or third year is a reasonable seed year. There are normally about 29,000 seeds/kg of *T. cordata* and 7,500/kg of *T. platyphyllos* of which 70 per cent germinate. The seed can be collected in October and must be stratified for 18 months before sowing.

Timber

The wood of limes is soft, light (about 560 kg/m^3 at 15 per cent moisture content), white when freshly cut and fine-textured. Its main uses are based on its softness and ability to resist splitting. It is easy to machine and is used in turnery and carving. In the Soviet Union it is used for furniture manufacture and many purposes for which plastics might be used in the west. From the middle ages to the 18th century the fibrous bark of lime was used for rope making.

TSUGA HETEROPHYLLA (Raf.) Sarg.

Western hemlock

Origin and introduction

There are ten species of *Tsuga* found in temperate parts of north America and eastern Asia. Western hemlock occurs naturally from south-east Alaska to south-east California and from sea level to 2,250 m. It is particularly abundant along the coasts of Washington and Oregon, especially on well drained soils in areas with mild, humid climates. It was introduced to Britain in 1851.

Climatic requirements

There are no marked preferences within Britain. *Tsuga* does well in the west and may be highly productive in quite low rainfall areas, but is less suited to the drier east and south-east than Douglas fir. It is mildly susceptible to late spring frosts, but recovers well. Though generally windfirm, in exposed areas, it is defoliated by strong winds and is more susceptible to dieback of the leading shoot than any other common conifer, especially when young. Its ability to grow replacement shoots can result in multiple-stemmed trees on both exposed and frost-prone sites.

Site requirements

Hemlock does best on deep, moist and well aerated soils but grows well on both wet and dry sites and the better peats. It thrives on moderately fertile soils at low elevations but will grow in wetter regions on infertile soils. It will not grow on limestone sites.

Figure 30. Western hemlock, *Tsuga heterophylla.*

Other silvicultural characteristics and yield

Tsuga is more shade-bearing than any other species commonly grown in Britain, and is easier to establish under shade than in the open. Natural regeneration is often very good, even in dense shade. It is difficult to establish pure on bare ground and does better with a nurse. Like the spruces, it does not compete well with heather. *Tsuga* is more susceptible than many species to attacks by *Heterobasidion annosum* and *Armillaria* spp., and infected sites should be avoided.

According to Aldhous and Low (1974), on financial grounds, hemlock is often second choice to Sitka spruce and grand fir. It could have a place on soils too infertile for *Abies grandis* and too dry for Sitka spruce, but in southern Britain it again comes second, this time to Corsican pine. In east Scotland and north-east England it is a good alternative to Scots and lodgepole pines if shelter is adequate.

Yield classes range from 12 to 24 m^3/ha/year.

Flowering, seed production and nursery conditions

The tree flowers in April; seeds ripen, and can be collected in September, and are dispersed naturally from October to May. The earliest age at which the tree bears seeds is 20 to 30 years but the best seed crops are usually at three yearly intervals between 40 and 60 years. There are approximately 573,200 seeds/kg (range 416,700 - 1,119,900). Cones should be collected as soon as they change from bright green to yellow, and the tips of the seed wings are visible and a light brown colour. In nurseries, seed should normally be sown between late February and mid March. It needs no special treatment before sowing according to Aldhous (1972).

Provenance

Vancouver Island sources are recommended by Lines (1987) as good general purpose origins.

Timber

The timber is usually described as being intermediate between Douglas fir and spruce. It is non-resinous, white and typically straight grained. It has a high moisture content when green and must be seasoned slowly to avoid distortion. Like spruces, it is difficult to get preservatives to penetrate the wood. Western hemlock is used for constructional work, in situations where its relatively low strength and decay-prone timber is acceptable, and for interior woodwork and other purposes where a fairly high grade of softwood is needed. At 15 per cent moisture content, its density is about 500 kg/m^3.

ULMUS L. Elm

Interest in elms today is, unfortunately, largely academic because, since the advent of the virulent strain of Dutch elm disease in the 1970s, it has been hopeless to plant the species with any prospect of success. Young suckers remain very common in hedgerows but once they reach diameters of about 10 cm the bark becomes thick enough for beetles to breed under, and they are soon attacked and die. The disease has caused a profound change to the character of the English midlands in particular, where elms were once a notable feature of the agricultural landscape.

There are about 18 north temperate species of elms. Several species, sub-species and many hybrids are native to Britain. Apart from the distinct wych elm, *Ulmus glabra*, their classification is incompletely worked out and difficult to the person dealing with the trees in the field. Most are probably *U. procera* of which many strains have been planted over the centuries in different parts of the country. Some are recognised as separate species or sub species. Because of the ease by which *U. procera* can be propagated from suckers it is certain that within many localities clonal material has been used quite extensively in planting on fertile, generally heavy lowland soils. Wych elm, *U. glabra*, is the other main species. It is more a woodland tree of the north and west and does not produce suckers.

The English elm, *U. procera*, was very widely planted when hawthorn hedgerows were being established during the enclosure movements from 1550 to 1850. They are particularly well adapted to hedges, being able to grow vigorously from suckers which spring in profusion from roots of felled trees. Elms are strong light-demanders and they are among the best trees at tolerating atmospheric pollution and salty winds from the sea.

Elm leaves were, and still are used in some countries, as fodder for many domestic animals, and were lopped or "shredded" for this purpose.

Work on the selection and breeding of elms resistant to Dutch elm disease has been carried out since the 1930s, especially in the Netherlands (Burdekin and Rushforth, 1981). This approach is showing some promise and a few relatively resistant clones have been produced, but much still needs to be done before they can be planted with any confidence. They flower in early spring, before the leaves come out.

Galls on elm leaves are caused by several species of aphids some of which migrate in summer to the roots of economic plants and cause damage: *Schizoneura lanuginosa* to the roots of pear trees, *S. ulmi* to currants and gooseberries and *Tetraneura ulmi* to grasses and cereals (Blackman, 1974).

Timber

The wood of elm is notable for resisting strains that cause other timbers to split. Its uses arose mostly from this attribute. Lengths were once hollowed out to make water pipes (Edlin, 1965), and it was used for the seats of chairs, the hubs of wooden wheels and other similar purposes. The average density of the wood of *U. glabra* at 15 per cent moisture content is about 690 kg/m^3. *U. procera* is lighter, at 560 kg/m^3.

Figure 31. A) English elm, *Ulmus procera*; B) Wych elm, *Ulmus glabra*.

REFERENCES

Aldhous, J.R. (1972). Nursery practice. Forestry Commission *Bulletin No.* 43. 184 pp.

Aldhous, J.R. and Low, A.J. (1974). The potential of western hemlock, western red cedar, grand fir and noble fir in Britain. Forestry Commission *Bulletin No.* 49. 105 pp.

Anderson, M.L. (1950). *Selection of tree species.* Oliver and Boyd. 151 pp.

Anon (1980). Poplars. *Forestry and British Timber*, June 1980.

Arboricultural Association (1991). *Arboricultural Handout No. 4*, Tree Management. Arboricultural Association, Ampfield, Romsey, Hampshire. 2 pp.

Bastide, J.G.A. la , and Faber, P.J. (1972). Revised yield tables for six tree species in the Netherlands. *Uitvoerig Verslag, Stitching Bosboywproefstation "De Dorschkamp" 11* (1). 36 pp.

Batchelor (1924). Walnut culture in California. College of Agriculture, Agricultural Experimental Station, Berkeley, *Bulletin No. 379.*

Beaton, A. (1987). Poplars and agroforestry. *Quarterly Journal of Forestry 81*, 225-233.

Becker, M., Tacon, F. le and Picard, J.F. (1978). Regeneration naturelle du hêtre et travail du sol. In *IUFRO symposium on the establishment and treatment of high quality hardwood forests in the temperate climatic region.* Nancy, France.

Begley, D.C. (1955). Growth and yield of sweet chestnut coppice. Forestry Commission *Forest Record No. 30.* 25 pp.

Bevan, D. (1962). Pine shoot beetles. Forestry Commission *Leaflet No. 3.* 8 pp.

Bevan D. and Brown, R.M. (1978). Pine looper moth. Forestry Commission *Forest Record No. 119.* 13 pp.

Blackman, R. (1974). *Aphids.* London, UK, Ginn and Co. Ltd. 175 pp.

Brown, J.M.B. (1953). Studies on British beechwoods. Forestry Commission *Bulletin No. 20.* 100 pp.

Burdekin, D.A. and Rushforth, K.D. (1981). Breeding elms resistant to Dutch elm Disease. Department of Environment *Arboriculture Research Note No. 2/81.*

Cannell, M.G.R. and Smith, R.I. (1984). Spring frost damage on young *Picea sitchensis. Forestry 57*, 177-197.

Cannell, M.G.R., Sheppard, L.J., Smith, R.I. and Murray, M.B. (1985). Autumn frost damage on young *Picea sitchensis. Forestry 58*, 145-166.

Carey, M.L. and Barry, T.A. (1975). Coniferous growth and rooting patterns on machine sod-peat bog. *Irish Forestry* 32, 18-29.

Carter, C.I. (1977). Impact of green spruce aphid on growth. Forestry Commission *Research and Development Paper No. 116.* 8 pp.

Carter, C.I. and Nichols, J.F.A. (1988). The green spruce aphid and Sitka spruce provenances in Britain. Forestry Commission *Occasional Paper No. 19.* 7 pp.

Christie, J.M., Miller, A.C. and Brumm, L.E. (1974). *Nothofagus* yield tables. Forestry Commission *Research and Development Paper No. 106.* 5 pp.
Clapham, A.R., Tutin, T.G. and Warburg, E.F. (1985). *Excursion Flora of the British Isles.* Cambridge University Press. 499 pp.
Cobbett, W. (1829). *The English gardener* (1980 edition). Oxford University Press. 335 pp.
Critchfield, W.B. (1957). Geographic variation in *Pinus contorta.* Maria Moors Cabat Foundation *Publication 3.* Harvard University Press.
Davies, E.J.M. (1980). Useless? The case against contorta. *Scottish Forestry,* 34, 110-113 and letter on p. 156.
Dawkins, H.C. (1963). Crown diameters: their relation to bole diameter in tropical forest trees. *Commonwealth Forestry Review 42*, 318-333.
Day, W.R. (1951). Studies on the dying of spruces and depth of rooting in relation to root disease and butt rot. Forestry Commission *Research Branch Paper No. 4.* 53 pp.
Donoso, C. (1979). Genecological differentiation in *Nothofagus obliqua* in Chile. *Forest Ecology and Management* 2, 53-66.
Edlin, H.L. (1965). A modern sylva: 12. Elms. *Quarterly Journal of Forestry 59*, 41-51.
Edlin, H.L. (1966). A modern sylva: 17. Sequoias and their kin. *Quarterly Journal of Forestry 60*, 101-109.
Elias, T.S. (1980). *The complete trees of north America.* Van Nostrand Reinhold Co., New York. 948 pp.
Ellis, J. (1768). In *Philosophical Transactions of the Royal Society 58*, 75-78.
Elwes, H.J. and Henry, A. (1906). *The trees of Great Britain and Ireland.* Edinburgh. 8 volumes.
Engler, J.M., Louarn, H. le and Tacon, F. le (1978). Influence of birds and small rodents on beech nuts disappearing during winter. In *IUFRO symposium on establishment and treatment of high quality hardwood forests in the temperate climatic region.* Nancy, France. p. 77.
Evans, J. (1981). Broadleaves - silviculture. Forestry Commission *Report on Forest Research 1981.* pp 13-14.
Evans, J. (1982). Sweet chestnut coppice. Forestry Commission *Research Information Note No. 70/82.* 4 pp.
Evans, J. (1984). Silviculture of broadleaved woodland. Forestry Commission *Bulletin No. 62.* 232 pp.
Evans, J. (1986). A re-assessment of cold-hardy eucalypts in Great Britain. *Forestry 59*, 223-240.
Evans, J. and Fourt, D.F. (1981). Oak on heavy clays. Forestry Commission *Report on Forest Research 1981*, p. 14.
Evans, J., Baker, K.F., Preston, R.E. and Warn, R.E. (1982). Broadleaves - silviculture. Forestry Commission *Report on Forest Research 1982.* 7 pp.
Evelyn, J. (1678). *Sylva, or a discourse on forest trees.* London.

Everard, J.E. and Fourt, D.F. (1974). Monterey pine and bishop pine as plantation trees in southern Britain. *Quarterly Journal of Forestry 68*, 111-125.

Ferguson, T.P. and Bond, G. (1953). Observations on the formation and functions of the root nodules of *Alnus glutinosa*. *Annals of Botany (NS) 17*, 175-188.

Forestry Commission (1956). Utilisation of hazel coppice. Forestry Commission *Bulletin 27*, HMSO, London. 33 pp.

Forestry Commission (1961). *Megastigmus* flies attacking conifer seed. Forestry Commission *Leaflet No. 8*. 10 pp.

Forestry Commission (1963). Poplar cultivation. Forestry Commission *Leaflet No. 27*. 6 pp.

Forestry Commission (1967). *Keithia* disease of *Thuja plicata*. Forestry Commission *Leaflet No. 43*. 7 pp.

Fourt, D.F., Donald, D.G.M. and Jeffers, J.N.R. (1971). Corsican pine in southern Britain. *Forestry 44*, 189-207.

FPRL (1964). Tests on the timber of home-grown red oak. *Quarterly Journal of Forestry 58*, 55-61.

Fraser, A.I. (1966). Current Forestry Commission root investigations. *Forestry* (supplement), 88-93.

Franklin, J.F., Sorensen, F.C. and Campbell, R.K. (1978). Summerization of the ecology and genetics of the noble and California red fir complex. In *Proceedings of the IUFRO joint meeting of working parties*. Vancouver, Canada. Volume I, 133-139.

Gale, A.W. (1962). *Sequoia sempervirens*, its establishment and uses in Great Britain. *Quarterly Journal of Forestry 56*, 126-137.

Garfitt, J.E. (1989). Growing superior ash. *Quarterly Journal of Forestry 83*, 226-228.

Gerard, J. (1597). *The herbal*. London.

Green, R.G. (1957). *Pinus radiata* in Great Britain. *Australian Forestry 21*, 66-69.

Greenaway, W., English, S., Whatley, F.R. and Rood, S.B. (1991). Interrelationships of poplars in a hybrid swarm as studied by gas chromatography - mass spectrometry. *Canadian Journal of Botany 69*.

Gibbs, J.N. (1987). New innings for willow? Bid to stamp out "watermark" disease in Britain. *Forestry and British Timber*, June 1987, pp. 16 and 19.

Handley, W.R.C. (1963). Mycorrhizal associations and *Calluna* heathland. Forestry Commission *Bulletin* 36, 70 pp.

Hartesveldt, R.J., Harvey, H.T. and Stecker, R.E. (1975). The giant *Sequoia* of Sierra Nevada. US Department of the Interior, *National Park Service Publication No. NPS 120*. 180 pp.

Harvey, H.T., Shellhammer, H.S. and Stecker, R.E. (1980). Giant *Sequoia* ecology. US Department of the Interior, *National Park Service, Scientific Monograph Series No. 12*. 182 pp.

Helliwell, D.R. (1982). Silviculture of ash in Wessex. *Quarterly Journal of Forestry 76*, 103-108.

Henman, G.S. (1984). Oak wood structure and the problem of shake. In *Report of 4th meeting of National Hardwoods programme.* Commonwealth Forestry Institute, Oxford, pp. 10-16.

Hibberd, B.G. (1986). Forestry practice. Forestry Commission *Bulletin No. 14.* 104 pp.

Hibberd, B.G. (1988). Farm woodland practice. *Forestry Commission Handbook 3.* HMSO, London. 106 pp.

Hummel, F.C., Palz, W. and Grassi, G. (1988). *Biomass forestry in Europe: a strategy for the future.* Elsevier applied science publishers, Essex, UK. 600 pp.

Jarvis, P.G. (1964). The adaptability to light intensity of seedlings of *Quercus petraea. Journal of Ecology 52*, 545-571.

Jeffers, J.N.R. (1956). The yield of hazel coppice. In Forestry Commission *Bulletin 27*, 12-18.

Jobling, J. (1990). Poplars for wood production and amenity. Forestry Commission *Bulletin 92*, HMSO, London. 84 pp.

Jokela, J.J. and Sawtelle, R.A. (1985). Origin of oak stands on the Springfield plain: a lesson in oak regeneration. In *Proceedings of 5th Central Hardwood Conference, Department of Forestry, University of Illinois.* (Editors J.O. Dawson and K.A. Majerus), 181-188.

Jones, E.W. (1945). Biological flora of the British Isles, *Acer* L. *Journal of Ecology 32*, 215-252.

Jones, E.W. (1952). Natural regeneration of beech abroad and in England. *Quarterly Journal of Forestry 46*, 75-82.

Jones, E.W. (1954). In Review on studies on British beechwoods by J.M.B. Brown. *Forestry 27*, 152-155.

Jones, E.W. (1959). Biological flora of the British Isles, *Quercus* L. *Journal of Ecology 47*, 169-222.

Jones, F.G.W. and Jones, M.G. (1974). *Pests of field crops.* Edward Arnold, London. 448 pp.

Justice, J. (1759). A dissertation on the culture of forest trees. University of Edinburgh, Forestry Department *Bulletin No. 6*, (1959). 36 pp.

Kanowski, P.J., Mather, R.A. and Savill, P.S. (in press). Short note: Genetic control of oak shake; some preliminary results. *Silvae Genetica.*

Kennedy, D. (1985). The potential for genetic improvement of birch (*B.pendula*). In *Proceedings of the National Hardwoods Programme.* Commonwealth Forestry Institute, Oxford, pp. 2-7.

Kennedy, J.N. (1974). Selection of conifer seed for British forestry. Forestry Commission *Leaflet No. 60.* 6 pp.

Kent, N. (1779). *Hints to gentlemen of landed property.* London.

Keresztesi, B. (1983). Breeding and cultivation of black locust in Hungary. *Forest Ecology and Management 6*, 217-244.

Klemp, C.D. (1979). [Silvicultural establishment of walnut for high quality timber production]. *Allgemeine Forstzeitschrift No. 27*, 732-733.

Lake (1913). The Persian walnut industry of the United States. US Department of Agriculture *Bulletin No. 254*.

Lanier, L., Rameau, J.-C., Keller, R., Joly, H.-I., Drapier, N., and Sevrin, E. (1990). L'Alisier torminal. *Revue Forestiere Française 42*, 13-34.

Larsen, C.S. (1946). [The flowering and breeding of ash]. *Forestry Abstracts 47*, p. 22.

Laurent, C., Rondeux, J. and Thill, A. (1988). Production du chene rouge d'Amerique en moyenne et haute Belgique. Faculté des Sciences Agronomiques Gembloux, Belgique *Document D/1988/5117/02*. 37 pp.

Leather, S.R. (1987). Lodgepole pine seed origin and the pine beauty moth. Forestry Commission *Research Information Note 125*. 2 pp.

Lines, R. (1978a). The IUFRO experiments with *Abies grandis* in Britain: nursery stage. In *Proceedings of the IUFRO joint meeting of working parties*. Vancouver, Canada. Ministry of Forests, Victoria, B.C., Canada. Volume 2, 339-345.

Lines, R. (1978b). *Abies grandis*. Forestry Commission *Report on Forest Research 1978*. pp. 16-18.

Lines, R. (1979a). The IUFRO experiments with *Abies grandis* in Britain: nursery stage. In *Proceedings of the IUFRO joint meeting of working parties*. Vancouver, Canada. Ministry of Forests, Victoria, B.C., Canada. Volume 2, 339-345.

Lines, R. (1979b). The IUFRO experiments with Sitka spruce in Great Britain. In *Proceedings of the IUFRO joint meeting of working parties*. Vancouver, Canada. Volume 2, 211-224.

Lines, R. (1979c). Natural variation within and between the silver firs. *Scottish Forestry 33*, 89-101.

Lines, R. (1980). Species trials. Forestry Commission *Report on Forest Research 1980*, pp. 17-19.

Lines, R. (1985a). Species. Forestry Commission *Report on Forest Research 1985*, p. 13.

Lines, R. (1985b). The Macedonian pine in the Balkans and Great Britain. *Forestry 58*, 27-40.

Lines, R. (1986). Species. Forestry Commission *Report on Forest Research 1986*, p. 14.

Lines, R. (1987). Choice of seed origin for the main forest species in Britain. Forestry Commission *Bulletin No. 66*. 61 pp.

Lines, R. and Gordon, A.G. (1980). Choosing European larch seed origins for use in Britain. Forestry Commission *Research Information Note No. 57/80*. 4 pp.

Lines, R. and Potter, M.J. (1985). *Nothofagus*. Forestry Commission *Report on Forest Research 1985*, p. 20.

Linnard, S. (1987). The fate of beech nuts. *Quarterly Journal of Forestry 81*, 37-41.

Locke, G.M.L. (1978). The growing stock of regions. *Forestry 51*, 5-8.
Locke, G.M.L. (1987). Census of woodlands and trees 1979-1982. Forestry Commission *Bulletin No. 63*. 51 pp.
Lonsdale, D. and Wainhouse, D. (1987). Beech bark disease. Forestry Commission *Bulletin No. 69*. 15 pp.
Lorrain-Smith, R. and Worrell, R. (eds) (1991). *The commercial potential of birch in Scotland.* The Forestry Industry Committee of Great Britain, London. 91 pp.
Mabberley, D.J. (1990). *The plant book.* Cambridge University press. 707 pp.
Macdonald, J., Wood, R.F., Edwards, M.V. and Aldhous, J.R. (1957). Exotic forest trees in Great Britain. Forestry Commission *Bulletin No. 30*. 167 pp.
McNeill, J.D., Hollingsworth, M.K., Mason, W.L., Moffat, A.J., Sheppard, L.J.and Wheeler, C.T. (1989). Inoculation of *Alnus rubra* seedlings to improve growth and forest performance. Forestry Commission *Research Information Note 144*. 4 pp.
McVean, D.N. (1953a). Regional variation of *Alnus glutinosa* in Britain. *Watsonia 3*, 26-32.
McVean, D.N. (1953b). Biological flora of the British Isles: *Alnus*. *Journal of Ecology 41*, 447-466.
MAFF (1980). Bacterial canker of cherry and plum. Ministry of Agriculture, Fisheries and Food *Leaflet No. 592*. 6 pp.
Marshall, Mr (1803). *On planting and rural ornament: a practical treatise.* (3rd edition). Volumes 1, 408 pp and 2, 454 pp. Printed by W. Bulmer and Co., London.
Matthews, J.D. (1963). Factors affecting the production of seed by forest trees. *Forestry Abstracts* (review article), *24 (1)*, i-xii.
Matthews, J.D. (1987). The silviculture of alders in Great Britain. In Oxford Forestry Institute *Occasional Papers 34*, 29-38.
Menzies, M.I. and Chavasse, C.G.R. (1982). Establishment trials on frost-prone sites. *New Zealand Journal of Forestry 27*, 33-49.
Mercer, P.C. (1984). The effect on beech of bark-stripping by grey squirrels. *Forestry 57*, 199-203.
Meyer (1980). [Growing *Sorbus torminalis* in Forest District Grohnde]. *Nieder Sachsische Landesforstverwaltung No. 33*, 184-193.
Miles, J. (1980). Effects of trees on soil. In *Forest and woodland ecology.* Last, F.T. and Gardiner, A.S.(editors). Institute of Terrestrial Ecology, Cambridge, 85-88.
Mitchell, A.F. (1978). *A field guide to the trees of Britain and northern Europe.* Collins. 416 pp.
Mitchell, A.F. (1985a). Conifers. *Forestry Commission Booklet 15*. HMSO, London. 67 pp.
Mitchell, A.F. (1985b). Broadleaves. *Forestry Commission Booklet 20*. HMSO, London. 104 pp.

Mitchell, A. and Jobling, J. (1984). *Decorative trees for country, town and garden*. HMSO, London. 146 pp.

Morris, M.G. (1974). Oak as a habitat for insect life. Morris, M.G. and Perring, F.H. (editors). *The British oak*. Botanical Society of the British Isles, 274-297.

Morris, M.G. and Perring, F.H. (editors) (1974). *The British oak*. Botanical Society of the British Isles. 376 pp.

Murray, M.B., Cannell, M.G.R. and Sheppard, L.J. (1986). Frost hardiness of *Nothofagus procera* and *Nothofagus obliqua* in Britain. *Forestry 59*, 209-222.

Newbold, A.J. and Goldsmith, F.B. (1981). The regeneration of oak and beech: a literature review. *Discussion papers in conservation No. 33*. University College London. 112 pp.

Nicholls, P.H. (1981). Spatial analysis of forest growth. Forestry Commission *Occasional Paper No. 12*. 97 pp.

O'Carroll, N. (1978). The nursing of Sitka spruce: 1. Japanese larch. *Irish Forestry 35*, 60-65.

O'Driscoll, J. (1980). The importance of lodgepole pine in Irish forestry. *Irish Forestry 37*, 7-22.

Patterson, D. (1988). *Commercial timbers of the world* (fifth edition). Gower Technical Press, England. 339 pp.

Pawsey, R.G. (1964). Resin top disease of Scots pine. Forestry Commission *Leaflet No. 49*. 8 pp.

Pawsey, R.G. and Young, C.W.T. (1969). A reappraisal of canker and dieback of European larch. *Forestry 42*, 154-164.

Pearce, M.L. (1978). Provenance of Douglas fir. Forestry Commission *Report on Forest Research 1978*, p. 11.

Pearce, M.L. (1979). Provenance of Norway spruce. Forestry Commission *Report on Forest Research 1979*, p. 11.

Peterken, G.F. (1981). *Woodland conservation and management*. Chapman and Hall, London. 328 pp.

Pigott, D. (1988). The ecology and silviculture of limes. Oxford Forestry Institute *Occasional Papers 37*, 27-32.

Popov, S. (1981). [Morphological features and growth of the root system of walnut in relation to the methods of plantation establishment and tending]. *Gorskostopanska Nauka 18*, 25-33.

Potter, C.J., Nixon, C.J. and Gibbs, J.N. (1990). The introduction of improved poplar clones. *Quarterly Journal of Forestry 84*, 261-264.

Potter, M.J. (1987). Provenance selection in *Nothofagus procera* and *N. obliqua*. Forestry Commission *Research Information Note No. 114/87/SILS*. 2 pp.

Preece, D.F. (1977). Watermark disease in willow. Forestry Commission *Leaflet No. 20*.

Pryor, S.N. (1985). The silviculture of wild cherry or gean. *Quarterly Journal of Forestry 79*, 95-109.

Pryor, S.N. (1988). The silviculture and yield of wild cherry. Forestry Commission *Bulletin No. 75*. 23 pp.

Pyatt, D.G. and Craven, M.M. (1979). Soil changes under even-aged plantations. *The ecology of even-aged forest plantations.* Ford, E.D., Malcolm, D.C. and Atterson, J. (editors). Institute of Terrestrial Ecology, Cambridge, pp. 369-386.

Rackham, O. (1980). *Ancient woodland: its history, vegetation and uses in England.* Arnold, London. 402 pp.

Read, D.J. (1967). *Brunchorstia* die-back of Corsican pine. Forestry Commission *Forest Record No. 61.* 6 pp.

Rebmann, (1912). Neuere Erfahrung uber de Anzucht eineiger Juglandeen. *Allgemeine Forst- und Jagd-Zeitung 88*, 257-272 and 401-403.

Reneaume, Monsieur (1700-1701). In *Philosophical Transactions of the Royal Society*, 908-911.

Roach, F.A. (1985). *Cultivated fruits of Britain.* Basil Blackwell, Oxford. 349 pp.

Rollinson, T.J.D. and Evans, J. (1987). The yield of sweet chestnut coppice. Forestry Commission *Bulletin No. 64.* 20 pp.

Savill, P.S. (1986). Anatomical characters in the wood of oak which predispose trees to shake. *Commonwealth Forestry Review 62*, 109-116.

Savill, P.S. and Evans, J. (1986). *Plantation silviculture in temperate regions.* Clarendon Press, Oxford. 246 pp.

Savill, P.S. and Mather, R.A. (1990). A possible indicator of shake in oak: relationship between flushing dates and vessel sizes. *Forestry 63*, 355-362.

Savill, P.S. and Sandels, A.J. (1983). The influence of early respacing on the wood density of Sitka spruce. *Forestry 56*, 109-120.

Savill, P.S. and Spilsbury, M.J. (1991). Growing oaks at closer spacing. *Forestry* 64.

Schlich, W. (1891). *A manual of forestry* (Volume 2). Bradbury, Agnew and Co., London. 351 pp.

Schmeling, W.K.B. von (1981). [Distribution and silviculture of Sorbus torminalis]. *Allgemeine Forstzeitschrift No. 9/10*, 209-211.

Schumucker, T. (1942). *Tree species.* Centre International de Sylviculture, Berlin-Wannsee.

Schwappach (1926). Ergebnisse Anbouversuche fremlandishen *Holzarten 37.*

Sheppard, L.J. and Cannell, M.G.R. (1987). Frost hardiness of subalpine eucalypts in Britain. *Forestry 60*, 239-248.

Southwood, T.R.E. (1961). The number of species of insect associated with various trees. *Journal of Animal Ecology 30*, 1-8.

Speight, M.R. and Wainhouse, D. (1989). *Ecology and management of forest insects.* Clarendon Press, Oxford. 374 pp.

Spence, H. and Witt, A.W. (1930). Walnuts. *Journal of the Royal Horticultural Society 60*, part 2, 244-265.

Steinhoff, R.J. (1978). Distribution, ecology, silvicultural characteristics and genetics of the *Abies grandis - Abies concolor* complex. *Proceedings of the IUFRO joint meeting of working parties.* Vancouver, Canada. Vol. I, 123-132.

Stevenson G.F. (1985). The silviculture of ash and sycamore. In *Proceedings of National Hardwoods Programme*. Commonwealth Forestry Institute, Oxford, pp. 25-31.

Stoakley, J.T. (1979). Pine beauty moth. Forestry Commission *Forest Record No. 120*. 11 pp.

Taylor, N.W. (1985). The sycamore in Britain - its natural history and value to wildlife. University College London, *Discussion Papers in Conservation 42*. 58 pp.

Theophrastus (c 200 BC). *Enquiry into plants and minor works on odours and weather signs*. Translated by A. Hort. Heinemann, London. Volume 1, 419 pp.

Thill, A. (1978). La sylviculture du frene en Belgique. In *IUFRO symposium on establishment and treatment of high quality hardwood forests in the temperate climatic region*, pp. 207-218.

Tuley, G. (1979). Fast growing pines. Forestry Commission *Report on Forest Research 1978*, p. 11.

Tuley , G. (1980). *Nothofagus* in Britain. Forestry Commission *Forest Record No. 122*. 75 pp.

Varty, I.W. (1956). *Adelges* insects of silver firs. Forestry Commission *Bulletin No. 26*. 75 pp.

Venables, R.G. (1985). The broadleaved markets. In *Growing timber for the market*. (Editor P.S. Savill). Institute of Chartered Foresters Edinburgh, pp. 22-29.

Walker, L.C. and Wiant, H.V. (1973). Silviculture of longleaf pine. Stephen F. Austin State University, Texas, *School of Forestry Bulletin No. 2*. 105 pp.

Wardle, P. (1961). Biological flora of the British Isles: *Fraxinus excelsior* L. *Journal of Ecology* 7, 173-203.

Warren-Wren, S.C. (1965). The significance of the caerulean or cricket bat willow. *Quarterly Journal of Forestry 59*, 193-205.

Watt, A.S. (1919). On the causes of failure of natural regeneration in British oakwoods. *Journal of Ecology* 7, 173-203.

Weissen, F. (1978). Dix années d'observations sur la régénération en hêtraie Ardennaise. In *IUFRO symposium on establishment and treatment of high quality hardwood forests in the temperate climatic region*. Nancy, France, pp. 60-70.

Wignall, T.A., Browning, G. and MacKenzie, K.A.D. (1985). Epicormic bud physiology and control. In *Proceedings National Hardwoods Programme 1985*, pp. 17-24.

Wigston, D.L. (1980). A preliminary investigation of the ecological implications of the introduction of species of *Nothofagus* into British forestry... UK; *Nature Conservancy Council*. 139 pp.

Wood, R.F. and Nimmo, M. (1962). Chalk downland afforestation. Forestry Commission *Bulletin No. 34*. 75 pp.

Young, C.W.T. (1978). Sooty bark disease of sycamore. Department of the Environment *Arboricultural Leaflet No. 3*. 8 pp.

APPENDIX

Crown diameter/bole diameter ratios

Most trees maintain an almost constant ratio of crown diameter to bole diameter throughout the silviculturally critical parts of their rotations. This is particularly the case with light demanding species. The ratio can slightly diminish with age, but has never been found to increase (Dawkins, 1963). Thus, if the ratio is 20, a healthy tree of mean diameter *d* must, on average, have a crown *20* x *d* across, and of course the same mean spacing between trees if the crown is not to become constricted. A knowledge of the ratio for individual species enables "thinning regimes" to be drawn up quite easily, which depend upon the mean diameter of stem to be achieved, rather than age.

Appendix Table 1 shows regressions of crown diameters on stem diameters at breast height for some of the more common species grown in the UK. In general, the species at the top of Table 1, whose crown diameters increase rapidly with increasing bole diameters, are among the most light demanding trees. Those near the bottom are more shade bearing.

Example: the expected crown diameter of ash (*Fraxinus excelsior*) for any mean stem diameter can be calculated as follows:

$$\text{Crown diameter (in metres)} = 0.7590 + (0.1890 \times \text{dbh})$$

So, if the dbh is 40 cm, the crown diameter will average 8.3 m, and the crown will occupy an area of 54.4 m^2, indicating that if competition between crowns is to be avoided at the time the stems reach 40 cm dbh, there should be 10,000/54.4 = 184 stems/ha. Similarly, if the trees are then thinned to a final crop spacing, and clear felled when the average dbh is 65 cm, the average diameter of the crowns will be:

$$\text{Crown diameter} = 0.7590 + (0.1890 \times 65) = 13.0 \text{ m}$$

The area occupied by a crown of 13 m diameter is 133.6 m^2, and therefore the number of trees remaining after thinning should be 10,000/133.6 = 75/ha, at an average spacing of 11.5 x 11.5 m.

Table 1. Regressions of crown diameters on stem diameters, listed in order of the size of the slopes of the regressions.

Species	No. of observations	Regression		r^2	dbh range
		slope	intercept		
Fraxinus excelsior	61	0.1890	0.7590	0.93	11 -113
Acer pseudoplatanus	51	0.1890	0.5930	0.82	14 - 91
Sorbus aucuparia	13	0.1845	1.0958	0.93	3 - 38
Quercus petraea and robur	62	0.1759	0.7728	0.95	10 -144
Larix decidua	72	0.1756	-0.0950	0.85	15 - 64
Larix kaempferi	31	0.1738	0.8467	0.97	6 - 57
Betula spp.	55	0.1619	0.9563	0.92	3 - 61
Fagus sylvatica	61	0.1573	0.9510	0.94	10 -149
Picea abies	50	0.1407	0.1056	0.88	6 - 50
Picea sitchensis	37	0.1352	0.8107	0.91	6- 78
Pinus nigra	42	0.1259	0.6803	0.90	7 - 82
Tilia cordata	33	0.1144	2.2593	0.91	10 - 75
Castanea sativa	43	0.1067	2.7915	0.76	16 - 93
Pseudotsuga menziesii	26	0.0991	2.1415	0.92	29 - 90
Thuja plicata	32	0.0824	1.4191	0.77	7 - 40

Life expectancy of some species

Although some trees can survive for several centuries, most species have a life expectancy of no more than 250 years in normal British conditions; some are unlikely to survive for more than 50 or 60 years before they start to die back or shed branches. Damage caused by poor pruning, impacts and soil compaction by vehicles, squirrels, and a variety of other causes can reduce this life span considerably. The Arboricultural Association (1991) has produced the following list of trees in six categories of useful, safe life expectancy, under garden or parkland conditions. Life expectancies in forest conditions might often be somewhat less:

Table 2. Life expectancy of trees.

Life expectancy (years)	Species
300 or more	Taxus baccata
200 to 300	Acer pseudoplatanus Castanea sativa Quercus petraea Quercus robur Tilia spp.
100 to 150	Fraxinus excelsior Juglans regia Picea abies Quercus rubra
70 to 100	Malus sylvestris Prunus avium Robinia pseudoacacia Sorbus aria Sorbus aucuparia
50 to 70	Alnus spp. Betula spp. most Populus spp. Salix spp.

Rotations for common species

The normal commercial felling age of any species is usually the age at which maximum net discounted revenue is obtained. This is determined by carrying out detailed economic analyses, incorporating such factors as discount rate, yield class, costs and revenues over the rotation and current market prices. Since detailed costs and revenues are seldom known for individual plantations, the calculations are normally based on current averages for the site and crop in question. Depending upon the assumptions which are made, the optimum felling age often occurs 10 to 15 years before the age of maximum mean annual increment among conifers. The difference tends to be small with crops where the age of maximum mean annual increment occurs relatively early, or where the discount rate assumed is low, but increases as the age of maximum mean annual increment increases. With most broadleaved species, the optimum felling age is often after the age of maximum mean annual increment because it is only after this stage that there is a big increase in the value of the timber, as it gets into sizes suitable for sawmilling. For example, economic oak rotations are commonly up to 40 years longer than the ages of maximum mean annual increment, and those for ash and sycamore, 10 to 20 years longer.

Table 3 shows the ages of maximum mean annual increment for the more common species grown in the UK for which yield tables exist. It may serve as a rough guide as to possible rotation lengths if appropriate additions or subtractions are made.

Table 3. Ages of maximum mean volume increment (from Hamilton and Christie, 1971, unless stated otherwise). Figures in ***bold and italics*** indicate average yield classes in different regions of Great Britain.

Species	Yield class (m^3/ha/year)										
	24	22	20	18	16	14	12	10	8	6	4
Abies grandis[1]	52	53	54	55	56	57	-	-	-	-	-
Chamaecyparis	58	60	63	65	67	69	72	-	-	-	-
Larix decidua	-	-	-	-	-	-	47	49	***52***	***56***	60
kaempferi and x eurolepis	-	-	-	-	-	41	42	***44***	***47***	50	56
Picea abies	-	63	65	67	69	72	***75***	***79***	84	90	-
sitchensis	46	48	50	52	55	***57***	***59***	***61***	64	66	-
Pinus contorta	-	-	-	-	-	54	***56***	***60***	***65***	***71***	80
nigra	-	-	53	54	55	57	***59***	***61***	64	70	-
sylvestris	-	-	-	-	-	66	69	***73***	***77***	82	89
Pseudotsuga menziesii	50	51	52	54	56	***58***	***61***	***64***	-	-	-
Thuja plicata	58	60	63	65	67	69	72	-	-	-	-

Tsuga heterophylla	53	56	59	63	67	73	79	-	-	-	-
Acer pseudoplatanus	-	-	-	-	-	-	40	41	43	45	49
Betula spp.	-	-	-	-	-	-	40	41	43	45	49
Fagus sylvatica	-	-	-	-	-	-	-	80	87	***96***	107
Fraxinus excelsior	-	-	-	-	-	-	40	41	43	***45***	49
Nothofagus spp.[2]	-	-	-	27	28	31	38	46	-	-	-
Populus spp.	-	-	-	-	-	35	36	37	38	39	42
Prunus avium[3]	-	-	-	-	-	-	-	67	***55***	-	-
Quercus petraea and robur coppice[4]	- -	- -	- -	- -	- -	- -	- -	- -	68 25	79 33	***90*** 43

[1] Unless stated otherwise, the figures in this Table are based on those in Hamilton and Christie (1971).
[2] From Tuley (1980).
[3] From Pryor, S.N. (1985). An evaluation of silvicultural options for broadleaved woodland. *DPhil Thesis (unpublished), University of Oxford.* 247 pp.
[4] From Crockford, K.J. (1987). An evaluation of British woodlands for fuelwood and timber production. *DPhil Thesis (unpublished), University of Oxford.* 219 pp.

Maximum dimensions recorded by Mitchell (1978) for species in the United Kingdom

Table 4. Maximum dimensions of trees.

Scientific name	Maximum height (m)	Maximum dbh (m)
Table 4a. Important native timber producing species		
Fagus sylvatica	40	1.90
Fraxinus excelsior	45	1.90
Pinus sylvestris	35	1.60
Quercus petraea	30	2.70
robur	37	3.20
Table 4b. Minor native timber-producing species		
Alnus glutinosa	22	0.65
Betula pendula	30	0.95
pubescens	25	0.95
Carpinus betulus	30	1.30
Populus nigra var. betulifolia	35	1.60
Prunus avium	30	1.30
padus	15	-
Taxus baccata	25	-
Tilia cordata	32	1.90
platyphyllos	31	1.85

Table 4c. Other native trees		
Scientific name	Maximum height (m)	Maximum dbh (m)
Acer campestre	26	0.95
Corylus avellana	12	-
Ilex aquifolium	22	0.60
Malus sylvestris	10	0.95
Salix alba	25	-
caprea	16	0.30
fragilis	18	1.10
pentandra	10	-
Sorbus aria	20	0.60
aucuparia	20	0.80
torminalis	22	0.89
Ulmus glabra	38	2.25

Table 4d. Exotic conifers		
Scientific name	Maximum height (m)	Maximum dbh (m)
Abies alba	48	1.90
grandis	50	1.90
procera	47	1.65
Chamaecyparis lawsoniana	38	1.25
Cupressocyparis x leylandii	31	-
Larix decidua	45	1.75
kaempferi	32	0.80
x eurolepis	37	0.95
Picea abies	43	1.25
sitchensis	53	2.40
Pinus		
contorta	25	0.95
muricata	29	0.95
nigra var. nigra	33	1.25
nigra var. maritima	35	1.25
Pseudotsuga menziesii	55	2.25
Sequoia sempervirens	42	2.25
Sequoiadendron giganteum	50	2.01
Thuja plicata	41	1.60
Tsuga heterophylla	45	1.80

Table 4e. Exotic broadleaves		
Scientific name	Maximum height (m)	Maximum dbh (m)
Acer pseudoplatanus	35	2.20
Alnus incana	25	0.65
rubra	20	0.65
Castanea sativa	35	3.20
Eucalyptus gunnii	35	1.25
Juglans regia	23	0.95
Nothofagus obliqua	30	0.95
procera	26	0.95
Quercus rubra	35	1.90
cerris	38	2.55
Robinia pseudoacacia	30	1.60
Tilia x vulgaris	40	2.20

INDEX

Abies
 alba 4-5, 6, 7, 9, 101, 102, 137, 138
 grandis 5-8, 29, 118, 138
 procera 6-9, 62
Acer
 campestre 10, 11, 14, 137
 platanoides 10-12
 pseudoplatanus 2, 3, 10-15, 42, 43, 89, 114, 131, 132, 133, 135, 139
Adelges
 cooleyi (insect) 87
 piceae (insect) 7
Alder, *see Alnus*
Alnus
 cordata 16-17, 19
 glutinosa 17-21, 136
 incana 17-21, 139
 rubra 19, 21, 132, 139
Andricus
 quercuscalicis (insect) 89
Apodemus
 sylvaticus (wood mouse) 95
Armillaria
 spp (fungus) 18, 53, 118
Ash, *see Fraxinus excelsior*
Austrian pine, *see Pinus nigra var. nigra*

Beech, *see Fagus sylvatica*
Betula 22-24, 131, 132, 135, 136
 pendula 22-24, 136
 pubescens 22-24, 136
Bird cherry, *see Prunus padus*
Bishop pine, *see Pinus muricata*
Black alder, *see Alnus glutinosa*
Bupalis
 piniaria (insect) 72

Carpinus
 betulus 25-26, 48, 92, 114, 136
Castanea
 sativa 27-28, 58, 97, 131, 132, 139
Cavariella
 aegopodii (insect) 102
Chamaecyparis
 lawsoniana 29-30, 33, 138
 nootkatensis 33
Cherry, *see Prunus*
Clethrionomys
 glareolus (vole) 95
Coast redwood, *see Sequoia sempervirens*
Corsican pine, *see Pinus nigra var. maritima*
Corylus
 avellana 31-32, 137
Crab apple, *see Malus sylvestris*
Crimean pine, *see Pinus nigra var. caramanica*
Cronartium
 ribicola (fungus) 74, 76
Cryptococcus
 fagisuga (insect) 37
Cryptosoma
 corticale (fungus) 14
Cupressocyparis
 x leylandii 33, 138
Cupressus
 macrocarpa 33

Dendroctonus
 micans (insect) 63
Didymascella
 thujina (fungus) 111
Dothiciza
 spp (fungus) 81
Douglas fir, *see Pseudotsuga menzesii*
Downy birch, *see Betula pubescens*
Drepanosiphum
 platanoidis (insect) 14
Dreyfusia
 nüsslini (insect) 4
Dysaphis
 apiifolia (insect) 81

Elatobium
abietinum (insect) 63
Endothia
parasitica (fungus) 27
Erwinia
salicis (bacterium) 101
Eucallipterus
tiliae (insect) 115
Eucalyptus
debeuzevillei 35
gunnii 34, 35, 139
niphophila 35
European larch, *see Larix decidua*
European silver fir, *see Abies alba*

Fagus
sylvatica 1-3, 10, 13, 14, 16, 22, 26, 36-40, 42, 43, 45-48, 56-58, 73, 77, 90, 92, 95, 111, 114, 131, 135, 136
Field maple, *see Acer campestre*
Frankia
spp (bacterium) 18, 20
Fraxinus
excelsior 41-44, 130-132, 135, 136

Grand fir, *see Abies grandis*
Grey alder, *see Alnus incana*

Hazel, *see Corylus avellana*
Heterobasidion
annosum (fungus) 4, 7, 8, 29, 51, 53, 63, 66, 72, 87, 111, 118
Holly, *see Ilex aquifolium*
Hornbeam, *see Carpinus betulus*
Hybrid larch, *see Larix x eurolepis*

Ilex
aquifolium 44, 45
Italian alder, *see Alnus cordata*

Japanese larch, *see Larix kaempferi*
Juglans
regia 1, 46-49, 132, 139

Larix
decidua 3, 14, 50-54, 131, 134, 138
kaempferi 52-54, 58, 131, 134, 138
x eurolepis 3, 52-54, 134, 138

Lawson's cypress, *see Chamaecyparis lawsoniana*
Leyland cypress, *see Cupressocyparis leylandii*
Lime, *see Tilia*
Locust, *see Robinia pseudoacacia*
Lodgepole pine, *see Pinus contorta*

Macedonian pine, *see Pinus peuce*
Malus
baccata 55
pumila 55
sylvestris 55, 65, 70, 76, 132, 134, 136, 137
Marssonina
brunnea (fungus) 79, 81
Megacyllene
robiniae (insect) 99
Megastigmus
spermatrophus (insect) 87
Melampsora
spp (fungus) 79, 81, 102
Meria
laricis (fungus) 51
Monterey pine, *see Pinus radiata*
Mountain ash, *see Sorbus aucuparia*
Myzus
cerasi (insect) 83

Nectria
coccinea (fungus) 37
Noble fir, *see Abies procera*
Norway maple, *see Acer platanoides*
Norway spruce, *see Picea abies*
Nothofagus
obliqua 56-58, 139
procera 56-58,

Operophtera
brumata (insect) 94, 95
Oregon alder, *see Alnus rubra*

Panolis
flammea (insect) 66
Pemphigus
bursarius (insect) 79
phenax (insect) 79

Peridermium
pini (fungus) 77
Picea
abies 3-5, 10, 59-64, 96, 131, 132, 134, 138
obovata 59
sitchensis 2, 3, 5, 7, 8, 21, 24, 59-64, 66, 67, 73, 77, 86-88, 96, 103, 118, 131, 134, 138
Pinus
contorta 3, 62, 65-67, 69, 73, 134, 138
contorta ssp contorta 67
contorta ssp latifolia 66
contorta ssp murrayana 66
muricata 67-69, 138
nigra 68-73, 78, 80, 131, 134, 136, 138
nigra var. caramanica 68
nigra var. cebennensis 68
nigra var. maritima 3, 68, 69, 71-73, 77, 118
nigra var. nigra 69, 73, 138
palustris 74
peuce 1, 70, 73-74, 76,
radiata 68, 70, 75
strobus 70, 74, 76
sylvestris 2, 3, 24, 55, 65, 67, 70-72, 74, 76-77, 88, 96, 132, 134, 136, 137
sylvestris var. scotica 76
Poplar, *see Populus*
Populus 21, 58, 78-81, 99,
canescens 78, 80, 81
deltoides 78
nigra 78
nigra var. betulifolia 78, 80
tremula 78, 80, 81
trichocarpa 78
Poria
spp (fungus) 18
Prunus
avium 42, 48, 58, 82-85, 107, 111, 132, 136
cerasus 83, 85
padus 82, 136
Pseudomonas
juglandis (bacterium) 48
mors-prunorum (bacterium) 83
savastanoi (bacterium) 42
seringae (bacterium) 83
Pseudotsuga
menziesii 3, 7, 86-88, 111, 117, 118, 131, 134, 138
Pyrenean pine, *see Pinus nigra var. cebennensis*

Quercus
cerris 89, 93, 94, 139
imbricata 90
marilandica 90
petraea 89, 90-97, 131, 132, 135, 136
robur 89, 90-97, 131, 132, 135, 136
rubra 93, 97, 98, 139
stellata 90

Racinonis
buoliana (insect) 68
Radiata pine, *see Pinus radiata*
Red alder, *see Alnus rubra*
Rhizoctonia
solani (fungus) 38, 39
Rhopalosiphum
insertum (insect) 106
padi (insect) 83
Rhyacionia
buoliana (insect) 72, 75
Rhytisma
acerina (fungus) 14
Robinia
pseudoacacia 99-100, 132, 139
Rowan, *see Sorbus aucuparia*

Salix 100-102
alba 101, 102
aquatica gigantea 102
caprea 101, 137
fragilis 101, 102, 137
pentandra 101, 137
viminalis 101
x sepulcralis 101

Sceleroderrus
lagerbergii (fungus) 71
Schizolachnus
pineti (insect) 72
Schizoneura
lanuginosa (insect) 119
ulmi (insect) 119
Scots pine, *see Pinus sylvestris*
Sequoia
sempervirens 103-105, 138
Sequoiadendron
giganteum 104, 138
Silver birch, *see Betula pendula*
Sitka spruce, *see Picea sitchensis*
Sorbus
aria 106, 108, 132, 137
aucuparia 1, 13, 17, 22, 83, 106, 108, 131, 132, 137
torminalis 1, 106-108, 137
Sweet chestnut, *see Castanea sativa*
Sycamore, *see Acer pseudoplatanus*

Taphrina
betulina (fungus) 23
Taxus
baccata 109-110, 114, 132, 136
Tetraneura
ulmi (insect) 119
Thuja
plicata 10, 29, 111-113, 131, 134, 138
Tilia
cordata 1, 14, 114-116, 131, 136
platyphyllos 114-116, 136
x vulgaris 114, 116, 139
Tortrix
grossana (insect) 38
viridana (insect) 94, 95
Trichoscyphella
wilkommii (bacterium) 51
Tsuga
heterophylla 29, 117-118, 135, 138

Ulmus
glabra 119-120, 137
procera 119, 120

Walnut, *see Juglans*
Weeping willow, *see Salix x sepulcralis*
Western red cedar, *see Thuja plicata*
Weymouth pine, *see Pinus strobus*
Whitebeam, *see Sorbus aria*
Wild service tree, *see Sorbus torminalis*

Xanthomonas
populi (bacterium) 79

Yew, *see Taxus baccata*